Applications of Zero-Suppressed Decision Diagrams

Synthesis Lectures on Digital Circuits and Systems

Editor
Mitchell A. Thornton, *Southern Methodist University*

The *Synthesis Lectures on Digital Circuits and Systems* series is comprised of 50- to 100-page books targeted for audience members with a wide-ranging background. The Lectures include topics that are of interest to students, professionals, and researchers in the area of design and analysis of digital circuits and systems. Each Lecture is self-contained and focuses on the background information required to understand the subject matter and practical case studies that illustrate applications. The format of a Lecture is structured such that each will be devoted to a specific topic in digital circuits and systems rather than a larger overview of several topics such as that found in a comprehensive handbook. The Lectures cover both well-established areas as well as newly developed or emerging material in digital circuits and systems design and analysis.

Applications of Zero-Suppressed Decision Diagrams
Tsutomu Sasao and Jon T. Butler
2014

Modeling Digital Switching Circuits with Linear Algebra
Mitchell A. Thornton
2014

Arduino Microcontroller Processing for Everyone! Third Edition
Steven F. Barrett
2013

Boolean Differential Equations
Bernd Steinbach and Christian Posthoff
2013

Bad to the Bone: Crafting Electronic Systems with BeagleBone and BeagleBone Black
Steven F. Barrett and Jason Kridner
2013

Introduction to Noise-Resilient Computing
S.N. Yanushkevich, S. Kasai, G. Tangim, A.H. Tran, T. Mohamed, and V.P. Shmerko
2013

Atmel AVR Microcontroller Primer: Programming and Interfacing, Second Edition
Steven F. Barrett and Daniel J. Pack
2012

Representation of Multiple-Valued Logic Functions
Radomir S. Stankovic, Jaakko T. Astola, and Claudio Moraga
2012

Arduino Microcontroller: Processing for Everyone! Second Edition
Steven F. Barrett
2012

Advanced Circuit Simulation Using Multisim Workbench
David Báez-López, Félix E. Guerrero-Castro, and Ofelia Delfina Cervantes-Villagómez
2012

Circuit Analysis with Multisim
David Báez-López and Félix E. Guerrero-Castro
2011

Microcontroller Programming and Interfacing Texas Instruments MSP430, Part I
Steven F. Barrett and Daniel J. Pack
2011

Microcontroller Programming and Interfacing Texas Instruments MSP430, Part II
Steven F. Barrett and Daniel J. Pack
2011

Pragmatic Electrical Engineering: Systems and Instruments
William Eccles
2011

Pragmatic Electrical Engineering: Fundamentals
William Eccles
2011

Introduction to Embedded Systems: Using ANSI C and the Arduino Development Environment
David J. Russell
2010

Arduino Microcontroller: Processing for Everyone! Part II
Steven F. Barrett
2010

Arduino Microcontroller Processing for Everyone! Part I
Steven F. Barrett
2010

Digital System Verification: A Combined Formal Methods and Simulation Framework
Lun Li and Mitchell A. Thornton
2010

Progress in Applications of Boolean Functions
Tsutomu Sasao and Jon T. Butler
2009

Embedded Systems Design with the Atmel AVR Microcontroller: Part II
Steven F. Barrett
2009

Embedded Systems Design with the Atmel AVR Microcontroller: Part I
Steven F. Barrett
2009

Embedded Systems Interfacing for Engineers using the Freescale HCS08 Microcontroller
II: Digital and Analog Hardware Interfacing
Douglas H. Summerville
2009

Designing Asynchronous Circuits using NULL Convention Logic (NCL)
Scott C. Smith and JiaDi
2009

Embedded Systems Interfacing for Engineers using the Freescale HCS08 Microcontroller
I: Assembly Language Programming
Douglas H.Summerville
2009

Developing Embedded Software using DaVinci & OMAP Technology
B.I. (Raj) Pawate
2009

Mismatch and Noise in Modern IC Processes
Andrew Marshall
2009

Asynchronous Sequential Machine Design and Analysis: A Comprehensive Development
of the Design and Analysis of Clock-Independent State Machines and Systems
Richard F. Tinder
2009

An Introduction to Logic Circuit Testing
Parag K. Lala
2008

Pragmatic Power
William J. Eccles
2008

Multiple Valued Logic: Concepts and Representations
D. Michael Miller and Mitchell A. Thornton
2007

Finite State Machine Datapath Design, Optimization, and Implementation
Justin Davis and Robert Reese
2007

Atmel AVR Microcontroller Primer: Programming and Interfacing
Steven F. Barrett and Daniel J. Pack
2007

Pragmatic Logic
William J. Eccles
2007

PSpice for Filters and Transmission Lines
Paul Tobin
2007

PSpice for Digital Signal Processing
Paul Tobin
2007

PSpice for Analog Communications Engineering
Paul Tobin
2007

PSpice for Digital Communications Engineering
Paul Tobin
2007

PSpice for Circuit Theory and Electronic Devices
Paul Tobin
2007

Pragmatic Circuits: DC and Time Domain
William J. Eccles
2006

Pragmatic Circuits: Frequency Domain
William J. Eccles
2006

Pragmatic Circuits: Signals and Filters
William J. Eccles
2006

High-Speed Digital System Design
Justin Davis
2006

Introduction to Logic Synthesis using Verilog HDL
Robert B.Reese and Mitchell A.Thornton
2006

Microcontrollers Fundamentals for Engineers and Scientists
Steven F. Barrett and Daniel J. Pack
2006

Applications of Zero-Suppressed Decision Diagrams

Tsutomu Sasao and Jon T. Butler

ISBN: 978-3-031-79869-6 paperback
ISBN: 978-3-031-79870-2 ebook

DOI 10.1007/978-3-031-79870-2

A Publication in the Springer series
SYNTHESIS LECTURES ON DIGITAL CIRCUITS AND SYSTEMS

Lecture #45
Series Editor: Mitchell A. Thornton, *Southern Methodist University*
Series ISSN
Print 1932-3166   Electronic 1932-3174

Applications of Zero-Suppressed Decision Diagrams

Tsutomu Sasao
Meiji University

Jon T. Butler
Naval Postgraduate School

SYNTHESIS LECTURES ON DIGITAL CIRCUITS AND SYSTEMS #45

ABSTRACT

A zero-suppressed decision diagram (ZDD) is a data structure to represent objects that typically contain many zeros. Applications include combinatorial problems, such as graphs, circuits, faults, and data mining. This book consists of four chapters on the applications of ZDDs.

The first chapter by Alan Mishchenko introduces the ZDD. It compares ZDDs to BDDs, showing why a more compact representation is usually achieved in a ZDD. The focus is on sets of subsets and on sum-of-products (SOP) expressions. Methods to generate all the prime implicants (PIs), and to generate irredundant SOPs are shown. A list of papers on the applications of ZDDs is also presented. In the appendix, ZDD procedures in the CUDD package are described.

The second chapter by Tsutomu Sasao shows methods to generate PIs and irredundant SOPs using a divide and conquer method. This chapter helps the reader to understand the methods presented in the first chapter.

The third chapter by Shin-Ichi Minato introduces the "frontier-based" method that efficiently enumerates certain subsets of a graph.

The final chapter by Shinobu Nagayama shows a method to match strings of characters. This is important in routers, for example, where one must match the address information of an internet packet to the proper output port. It shows that ZDDs are more compact than BDDs in solving this important problem.

Each chapter contains exercises, and the appendix contains their solutions.

KEYWORDS

logic function, prime implicant, sum-of-products expression, binary decision diagram, zero-suppressed decision diagram, graph enumeration, CUDD package, frontier-based method, data-structure, non-deterministic automata, regular expression matching, one-hot code, intrusion detection

Contents

Preface

This book focuses on ZDDs or zero-suppressed decision diagrams. Originally called zero-suppressed BDDs, ZDDs have since developed into a research tool for combinatorial algorithms, including sets of combinations, symbolic logic, probability theory, and string matching for computer virus detection, text retrieval from databases, and DNA matching. D. E. Knuth has described ZDDs as *the most beautiful construct in computer science.*

This book is a tutorial description designed to educate a reader with no previous background in decision diagrams. While the target audience consists of graduate students, established researchers with little background in decision diagrams will benefit from the book's tutorial approach. The first chapter is based on a technical report by a well-known expert in logic synthesis. The last two chapters were selected from the papers presented at the Reed-Muller Workshop (RM-2013) on May 23–24, 2013, in Toyama, Japan. These chapters are rewritten for the book so that non-experts can understand them. Each chapter

1. is self-contained,

2. has examples and illustrations to explain the concepts presented,

3. introduces important previous work with simple examples, and

4. contains exercises for the reader to solve.

Structure of the book.

Preface: T. Sasao and J. T. Butler

Chapter 1: A. Mishchenko, "An introduction to zero-suppressed decision diagrams."

Chapter 2: T. Sasao, "Efficient generation of prime implicants and irredundant sum-of-products expressions."

Chapter 3: S. Minato, "The power of enumeration BDD/ZDD-based algorithms for tackling combinatorial explosion."

Chapter 4: S. Nagayama, "Regular expression matching using zero-suppressed decision diagrams."

Appendix: Solutions to the exercises.

Index

Authors' and Editors' Biographies

Acknowledgments

This work is supported in part by the Grants in Aid for Scientific Research of JSPS.

The authors and editors would like to thank many reviewers who provided detailed comments improving the chapters.

A preliminary version of this book was used as a textbook for a group seminar at Meiji University. Numerous improvements were proposed by the students of Meiji University: Yuta Chiku, Ichidou Fumishi, Yusuke Goto, Hikaru Kitagawa, Yuji Kitamura, and Atsushi Tsumuraya, assisted by Prof. Yukihiro Iguchi.

Prof. Yusuke Matsunaga's comments improved the readability of the manuscript.

Yuta Urano helped in the editing of this book.

Tsutomu Sasao, Kawasaki, Kanagawa, Japan
Jon T. Butler, Monterey, CA, USA
November 2014

CHAPTER 1

Introduction to Zero-Suppressed Decision Diagrams

Alan Mishchenko

CHAPTER SUMMARY

There is widespread use of zero-suppressed decision diagrams (ZDDs) in the design of logic circuits and in the efficient solution of combinatorial problems on sets, e.g., graph problems. ZDDs are especially efficient when there are relatively few elements compared to a large number of possibilities.

In this chapter, we focus on circuit design, including circuits based on the irredundant sum-of-products form. In addition to completely specified Boolean functions, this chapter also considers incompletely specified functions. Upon completion of this chapter, the reader will have an understanding of how to implement ZDD procedures on the popular decision diagram package CUDD [39]. It is assumed that the reader is familiar with Boolean algebra. For example, completion of an undergraduate course in logic design is sufficient preparation for this chapter [37].

1.1   INTRODUCTION

Since Bryant [4] put the theory of binary decision diagrams (BDDs) on a firm mathematical foundation in 1986, BDDs and BDD variants have found wide application in logic design. This has been especially true in computer-aided design (CAD). For example, BDDs have been used to develop minimal sum-of-products (SOP) expressions [11][40]. This has allowed logic designers to efficiently handle the vast logic resources available in modern VLSI circuits. It has transformed logic design into the specification of a logic function, leaving the design details to a computer implementation of a BDD.

Formal verification is typically a complex process of determining that a completed logic design meets the design specifications. Here also, BDDs have been effective [8]. BDDs are useful

A preliminary version of this chapter appeared as [31].

because they provide a *canonical* representation of a Boolean function. In a canonical representation, each Boolean function has only one representation. To illustrate, a sum-of-products expression is not canonical, since most functions are represented by more than one sum-of-products expression. For example, the sum-of-products expression $x_1 \vee x_2$ represents the same function as $x_1 \vee \bar{x}_1 x_2$. However, a truth table is a canonical representation of a Boolean function, since there is only one truth table for each function. In a similar way, BDDs provide a canonical representation of a Boolean function. This observation is important in formal verification, since it allows one to prove the equivalence of two Boolean functions by showing that their BDDs are equivalent.

In spite of their advantages, BDDs are not a panacea. In some applications, BDDs grow so large as to make traversal quite time consuming. Especially, this is true when the application involves sparse sets as represented by their characteristic function [3].

A set is *sparse* if the number of elements in it is much smaller than the number of elements that could possibly be in it. Cube covers, a representation of a Boolean function that is similar to the sum-of-products expression, typically represent sparse sets because the number of cubes in a typical cube cover is usually much less than the total number of possible cubes. The problem of accommodating the large BDDs associated with cube covers on sparse sets can be solved by using zero-suppressed decision diagrams (ZDDs) [22]. The process of forming a ZDD is identical with that of forming a BDD with one notable exception. This difference represents a significant change in how efficient these are in various applications. This is discussed in more detail later.

The impact on logic design of both ZDDs and BDDs is substantial. BDDs do well in the representation of functions, while ZDDs do well in the representation of covers. There are efficient procedures that perform conversions between ZDDs and BDDs. Together, ZDDs and BDDs solve many problems in logic design, including sum-of-product minimization [10], three-level minimization [28][38], and decomposition [19]. ZDDs also find use in applications outside logic design, including graph theory, polynomial manipulation, and regular expressions. See Section 1.8 for a list of applications where ZDDs have been used.

This chapter is a tutorial introduction to ZDDs. The goal is to focus on three types of applications:

- ZDDs for sets,
- ZDDs for cube covers,
- mixed BDDs/ZDDs for functions and cube covers.

To this end, we first discuss the basic principles and uses of ZDDs. In particular, Section 1.3 focuses on the main differences between BDDs and ZDDs in Boolean functions, sets, and cube covers.

In Section 1.4, we classify and discuss the elementary ZDD operators provided by a DD package. Next, we explore the generic structure of a DD-based recursive procedure, which is important for understanding the following sections.

Section 1.5 shows how ZDDs can be used to manipulate sets. The set-union operator is considered in an illustrative example. The complete source code of this operator in the CUDD package is included and explained. It assumes the reader is familiar with the C programming language.

Section 1.6 introduces the basics of ZDDs for manipulation of cube covers using the cover-product operator as an illustrative example.

Section 1.7 contains two important examples of mixed ZDD/BDD applications: generation of a ZDD representing all primes for a completely specified Boolean function given by a BDD, and computation of a ZDD representing an Irredundant Sum-of-Products of the incompletely specified Boolean function.

Section 1.8 provides a complete list of ZDD applications published to date, followed by conclusions in Section 1.9.

Two appendices contain annotated lists of ZDD-based procedures implemented in the CUDD package [39] and the EXTRA library [32].

Ideally, after completing the tutorial, the reader should be able to write his or her own ZDD-based procedures using the CUDD package.

1.2   DEFINITIONS

A *set* is a collection of elements from a finite domain. The descriptor "collection" means that only the presence of elements matters, not their order. Sets $\{a, b\}$ and $\{b, a\}$ are identical, while sets $\{a, b, c\}$ and $\{a, b\}$ are different. Sometimes it is convenient to assume that the elements are initially ordered and appear in the sets only in that order. Assuming that a precedes b in the order, both $\{a, b\}$ and $\{b, a\}$ are represented as $\{a, b\}$.

A *literal* is a Boolean variable or its negation, e.g., $a, \bar{b}$. A *product* is the AND of literals, e.g., $\bar{b}c$. A *cover* is a set of products. The *cardinality* of the cover is the number of products. A *complement* of the cover S is a cover T such that the union of S and T is a tautology, that is, the constant-1 Boolean function, and for any pair of products in S and T, there is no overlap.

Let $f : B^n \to B, B \in \{0, 1\}$, be a completely specified Boolean function (CSF). Let $F : B^n \to \{0, 1, -\}$ be an incompletely specified Boolean function (ISF) represented by two CSFs: the *on-set*, $f^1 = \{x | F(x) = 1\}$ and the *don't-care-set*, $f^{dc} = \{x | F(x) = -\}$.

A CSF can be represented by a set of products. This is known as a *sum-of-products* (SOP). An SOP is *irredundant* if no products or literals can be removed without changing the value of the represented function.

A *minterm* is the product that contains every variable. On the other hand, if a product has missing variables, such a literal can be seen as a sum of the positive literal and the negative literal. This leads to splitting the product into two smaller products. In general, if a product is missing k literals, it is equal to the OR of 2^k minterms, created because each missing variable corresponds two literals.

An n-variable Boolean function can be considered as a set of vertices[1] of an n-dimensional hypercube. In such a case, a product corresponds to a cube, and a minterm corresponds to a vertex.

---

[1]In this book, a node is used to denote a point of decision diagram, while a vertex is used to denote a point of a graph.

For example, Fig. 1.1 represents the three-variable function.

$$f = abc \vee a\bar{b}c \vee \bar{a}bc.$$

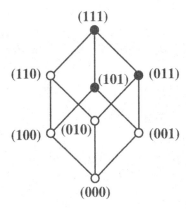

Figure 1.1: Three-variable function.

In Fig. 1.1, the vertex (111) corresponds to the minterm abc, and the vertex (011) corresponds to the minterm $\bar{a}bc$. Note that f can be simplified as $f = bc \vee ac$, where the product bc corresponds to the set {(111), (011)}.

The *area* of the Boolean space covered by the cube consists of all vertices created by splitting *don't-care* literals of the cube. Two areas of the Boolean space overlap if they have common vertices.

Two cubes are *disjoint* if the areas of the Boolean space covered by the cubes do not overlap. Two covers are *disjoint* if the areas covered by their cubes do not overlap.

Given a Boolean function f, the *negative cofactor* of f with respect to variable x is the Boolean function $f_{x=0} = f(x = 0)$. Similarly, the *positive cofactor* is the Boolean function $f_{x=1} = f(x = 1)$.

The following terminology is accepted in BDD research [4]. The BDD represents the function as a rooted directed acyclic graph. Each non-constant node N is labeled by a variable v and has edges directed toward two successor (children) nodes, else(N) and then(N), representing the cofactors of N with respect to v. Each constant node is labeled with 0 or 1. For a given assignment of the variables, the value of the function is found by tracing a path from the root to a constant node following the branches indicated by the values assigned to the variables. The function value is given by the constant node label.

Example 2.1 *Consider the four variable function*

$$F = ab \vee cd.$$

Fig. 1.2 shows the binary decision tree for F. The edges are directed downward. The dashed edges (solid

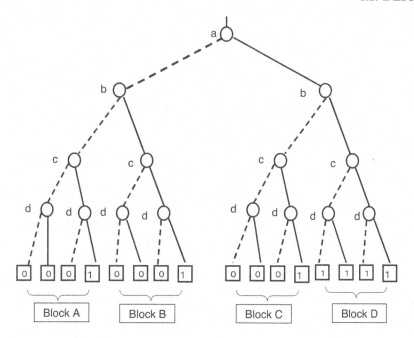

Figure 1.2: Decision tree for $F = ab \vee cd$.

edges) show that the value of the variables is 0 (1).

The total number of non-terminal nodes is

$$1 + 2 + 4 + 8 = 15.$$

The total number of terminal nodes is $2^4 = 16$, where 7 nodes have value 1, and 9 nodes have value 0.

(End of Example)

1.2.1 BDD AND ZDD REDUCTION RULES

To form a BDD or ZDD from the decision tree, we apply two reduction rules.

1. (BDD and ZDD) First, we merge equivalent subgraphs. Note that blocks A, B, and C are equivalent. In block A, in the subtree for $c = 0$, all the terminal nodes have the value 0. Also, in block D, all the terminal nodes have the value 1. So, we share them, resulting in the graph shown in Fig. 1.3.

In this graph, three subgraphs of blocks A, B, and C in Fig. 1.2 are shared. Also, in the block D, constant 1 terminals are shared.

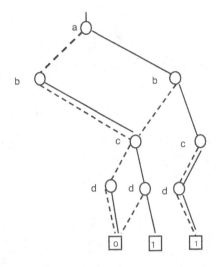

Figure 1.3: Decision diagram for $F = ab \vee cd$, after sharing equivalent subgraphs.

2. **(BDD)** To derive a BDD, we use the reduction rule for BDDs: Remove a node when both edges point the same node. Fig. 1.4 (a) shows the graph after applying the BDD reduction rule.

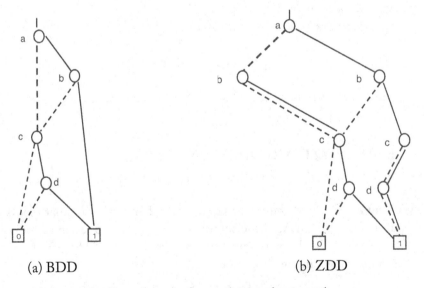

(a) BDD   (b) ZDD

Figure 1.4: Decision diagrams for $F = ab \vee cd$, after applying reduction rule.

2. (ZDD) To derive a ZDD, we use the reduction rule for ZDDs: Remove a node if its positive edge points to the constant-0 node. Fig. 1.4 (b) shows the graph after applying the ZDD reduction rule.

1.3   COMPARING BDDS AND ZDDS

Both BDDs and ZDDs can be seen as decision trees, simplified using two reduction rules that guarantee the canonicity of the representation. The first reduction rule (merging of equivalent subgraphs) holds for both BDDs and ZDDs; however, they differ in the second reduction rule (node elimination).

For BDDs, a node is removed from the decision tree if both its edges point to the same node. For ZDDs, the node is removed if its positive edge (then-edge) points to the constant-0 node. This variation in the rule, as mentioned before, explains the improved efficiency of ZDDs when handling sparse sets and the semantic differences between the two types of diagrams.

1.3.1   BOOLEAN FUNCTIONS

It can be shown that, in a BDD, all paths from the root to the constant-1 node represent cubes in a disjoint cover of the function. A variable is present in the *positive* (*negative*) polarity in the corresponding cube if the path contains the 1-edge (0-edge) of a node labeled by this variable; the variable is *absent* in the cube if the path does not go through a node labeled by this variable.

In a ZDD for the same function, all paths from the root to the constant-1 node also represent a disjoint cover of the function. (This cover is the same if the variable ordering is the same in both diagrams.) A variable is present in the *positive polarity* in the corresponding cube if the path goes through the 1-edge of a node labeled with this variable. A variable is present in the *negative polarity* in the cube if the path goes through the 0-edge or if the path does not go through a node labeled by this variable. A variable is *absent* in a cube, if the path goes through a node labeled by this variable and both edges of the node point to the same node.

Consider a BDD and a ZDD of the function $F = ab \lor cd$ shown in Fig. 1.5. Both the BDD and ZDD can be used to trace the disjoint cover of the function: $\{ab, \bar{a}cd, a\bar{b}cd\}$. As can be seen from Fig. 1.5, the size of the ZDD, expressed as the number of nodes in the diagram, is almost two times larger than that of the BDD. For this function, the ZDD is not as efficient as the BDD.

1.3.2   SETS OF SUBSETS

BDDs represent completely specified Boolean functions. In order to represent sets and sets of subsets, a set is put in one-to-one correspondence with its characteristic function [5].

Informally, the characteristic function of a set of subsets is a CSF that depends on as many input variables as there are elements that can potentially appear in a subset. For each subset in the set, one minterm appears in the on-set of the characteristic function. In this minterm, variables

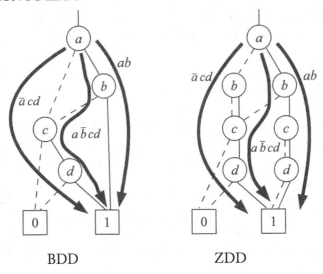

Figure 1.5: BDD and ZDD for $F = ab \vee cd$.

appear in positive (negative) polarities if they are present (absent) in the set. In the extreme case of one subset (that is, if a set of subsets contains only one subset), the characteristic function has only one minterm in its on-set.

For example, given three elements (a, b, c), consider the set of subsets $\{\{a, b\}, \{a, c\}, \{c\}\}$. If we associate each element with a binary variable having the same name, the characteristic function of the set of subsets is $F = ab\bar{c} \vee a\bar{b}c \vee \bar{a}\bar{b}c$. The first minterm corresponds to the subset $\{a, b\}$, the second to $\{a, c\}$, the third to $\{c\}$.

Important for our discussion are the following observations. The empty subset is represented by the minterm $F = \bar{a}\bar{b}\bar{c}$, while the subset containing all elements is represented by $F = abc$. The empty set is represented by the characteristic function $F = 0$, while the set of subsets composed of all possible subsets is represented by the characteristic function $F = 1$. The latter is obvious if we observe that the constant-0 function has no on-set minterms, while the constant-1 function has 2^n on-set minterms, corresponding to the complete Boolean space.

Note that there is a difference between the empty set of subsets and the set of subsets containing the empty set. The former has the characteristic function equal to the constant-0, while the latter has the characteristic function $F = \bar{a}\bar{b}\bar{c}$.

A characteristic function can be represented using a BDD or a ZDD. The two representations of the set of subsets $\{\{a, b\}, \{a, c\}, \{c\}\}$ are given in Fig. 1.6.

In both diagrams, there are three paths from the root node (on top) to the constant node 1 (at the bottom) corresponding to the subsets $\{a, b\}, \{a, c\}$, and $\{c\}$. The encoding of the variables is discussed in Section 3.1.

Note that the size of the ZDD in Fig. 1.6 is smaller than that of the BDD. It can be proved that an upper bound on the size of the ZDD is the total number of elements appearing in all subsets of the set of subsets. Meanwhile, an upper bound on the size of the BDD is given by the number of subsets multiplied by the number of all elements that can appear in them. This observation shows that ZDDs tend to be more compact when representing sets of subsets. The above theoretical upper bound on the ZDD size is rarely reached; in practice, ZDDs tend to be even more compact.

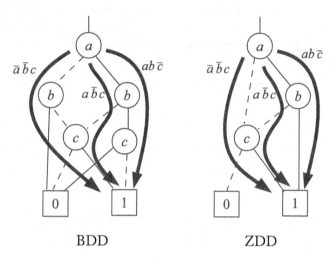

Figure 1.6: The BDD and the ZDD for the set of subsets $\{\{a, b\}, \{a, c\}, \{c\}\}$.

1.3.3   CUBE COVERS

Let us now consider the ZDD representation of a cube cover. First, it is necessary to introduce additional variables, because a BDD and a ZDD depending on the primary input variables represent a Boolean function and one disjoint cover of this function. To represent an arbitrary cover, each primary input is mapped into two variables: one of them represents the positive literal and another the negative literal. These variables are often kept adjacent in the variable order. Similarly to the set of subsets, a cube cover is represented by its characteristic function introduced as follows:

- The characteristic function of the cube cover depends on $2n$ variables representing literals, where n is the number of variables in the original Boolean function.
- For each cube of the cover, one minterm is added to the on-set of the characteristic function.
- The minterm has those variables in the positive polarity that correspond to literals present in the cube and those variables in the negative polarity that correspond to literals missing in the cube.

For example, to represent arbitrary covers of the four-variable function $F = ab \vee cd$, eight variables are used: $(a_1, a_0, b_1, b_0, c_1, c_0, d_1, d_0)$. These variables correspond to the positive and negative literals of each input variable.

Consider the cover $\{ab, cd\}$. The characteristic functions of this cover is: $\chi = a_1 \bar{a}_0 b_1 \bar{b}_0 \bar{c}_1 \bar{c}_0 \bar{d}_1 \bar{d}_0 \vee \bar{a}_1 \bar{a}_0 \bar{b}_1 \bar{b}_0 c_1 \bar{c}_0 d_1 \bar{d}_0$. The ZDD for the characteristic function is shown in Fig. 1.7.

Unlike the BDD for function χ, which depends on all eight variables, the ZDD depends on four variables only. These are variables that appear in the characteristic function in the positive or negative polarity and correspond to literals actually present in the cover. All other variables are missing in the ZDD, because according to the ZDD reduction rules a variable missing on a path is interpreted as a variable taking value 0 in the minterm of the characteristic function. This property makes ZDDs ideal for representing and manipulating large cube covers.

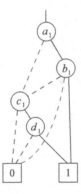

Figure 1.7: ZDD for the cover $\{ab, cd\}$.

The constant-0 node as a ZDD represents the empty cover. In this case, there are no assignments for which the characteristic function evaluates to 1, and therefore there are no cubes in the cover.

The constant-1 node represents the cover containing only the *tautology cube*, that is, the cube in which all the literals are missing. Indeed, there is only one path for which the characteristic function evaluates to 1, and this path does not go through any nodes. According to the ZDD reduction rules, it means that all the variables on the path are equal to 0, which in turn means that the cube contains no literals.

The decomposition of a cover with respect to a primary input variable is the triple of covers containing: (1) cubes with the variable as the positive literal, (2) cubes with the variable as the negative literal, (3) cubes without the variable (or with the variable as a *don't-care* literal).

The inverse operation is the composition of the three covers into one. The composition is performed using a variable that is not currently used in the covers. For example, consider the cover $C = ab\bar{c} \vee a\bar{b}d \vee ac \vee d$. Decomposing C with respect to the primary input variable b yields:

$C0 = ad$, $C1 = a\bar{c}$, $C2 = ac \vee d$. The reverse operation, the composition of the covers $C0$, $C1$, and $C2$ with respect to b, which does not appear in them, produces the initial cover C.

If the cover is represented by a ZDD, the decomposition and composition operations are performed by procedures of the ZDD package. In the traversals presented in this paper, the procedures are denoted *DecomposeCover()* and *ComposeCover()*. *DecomposeCover()* takes the cover and the primary input variable and returns three subcovers. *ComposeCover()* takes three subcovers and the primary input variable and returns the composed cover.

For further details on using ZDDs in the representation of cube covers, the reader is referred to [24][27][28], where some basic ZDD-based recursive operators are introduced and explained.

Here, the reader is encouraged to do Exercises 1.1, 1.2, 1.3, and 1.4.

1.4   BASIC ZDD PROCEDURES

The basic ZDD operators can be classified as follows:

1.4.1   PROCEDURES WORKING WITH FUNCTIONS

These procedures are similar to those developed to manipulate Boolean functions using BDDs.

- *Procedures returning elementary functions:*

  - Constant-0 ZDD (constant zero function, $F = 0$)
  - Universal ZDD (constant one function, $F = 1$)
  - Single-variable ZDD (the function equal to the elementary variable, $F = v$)

- *Procedures performing Boolean operations:*

  - If-Then-Else (ITE) operator. This function returns the result of applying ITE to A, B, and C:

$$ITE(A, B, C) = AB \vee \bar{A}C.$$

Note that the complement of a Boolean function is computed as follows: $\bar{F} = ITE(F, 0, 1)$.

1.4.2   PROCEDURES WORKING WITH COVERS

- *Procedures returning elementary sets:*

  - Constant-0 ZDD (the empty set, {})
  - Constant-1 ZDD (the set of subsets consisting of the empty set, {{}})
  - Single-variable ZDD (the set containing element v, {{v}})

- *Procedures performing operations on the set of subsets with respect to a single element (variable):*

  - $Subset0(S, v)$ returns the set of subsets of S not containing element v.
  - $Subset1(S, v)$ returns the set of subsets of S containing element v, and remove v.

○ *Change(S, v)* returns the set of subsets derived from S by adding element v to those subsets that did not contain it and removing element v from those subsets that contain it.

- *Procedures performing standard set operations for two sets of subsets:*

  ○ *Union(X, Y)* returns the set of subsets belonging to $X \cup Y$.
  ○ *Intersection(X, Y)* returns the set of subsets belonging to both $X \cap Y$.
  ○ *Difference(X, Y)* returns the set of subsets of X not belonging to Y: $X - Y$.

It is important to distinguish single-variable ZDDs, as used in the manipulation of Boolean functions, and those used in the manipulation of sets of subsets. Fig. 1.8 shows the ZDDs for $F = b$, assuming that there are three variables (a, b, c). In the case of functions, this ZDD represents function $F = b$. In the case of sets of subsets, the ZDD represents the set of subsets containing one set composed of a single element b. In this case, the characteristic function is $F = \bar{a}b\bar{c}$.

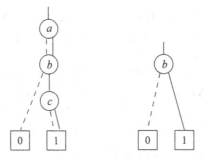

Figure 1.8: ZDDs for elementary variable b used in functions manipulation (left) and in set manipulation (right).

In addition to the above nine elementary operators defined for sets, the two pairs of set product and weak-division operators have been implemented using ZDDs. These two pairs of operators correspond to unate and binate algebras [24][28]. The remainder is another operator in each of these algebras. In practice, however, the remainder (%) is computed using product (*), weak-division (/), and set-difference (-) as follows: $X\%Y = X - Y * (X/Y)$.

Speaking informally, in unate algebra, every literal of all cubes is either missing or appears in only one polarity, while in binate algebra every literal may be positive, negative, or missing.

Correspondingly, to manipulate sets in unate algebra, every literal is encoded using one ZDD variable, while to manipulate sets in binate algebra, two literals are used. One represents the variable in positive polarity, the other represents it in negative polarity.[2]

---

[2]In the public distribution of the CUDD package, unate product and division are implemented as Cudd_zddUnateProduct() and Cudd_zddDivide(), while binate product and division are implemented as Cudd_zddProduct() and Cudd_zddWeakDiv().

In terms of the definitions introduced above, ZDDs used to manipulate sets of subsets implement the unate algebra, while ZDDs used to manipulate the cube covers implement the binate algebra.

1.4.3 GENERIC STRUCTURE OF A RECURSIVE ZDD PROCEDURE

In this subsection, we discuss the generic structure of a recursive ZDD procedure. The presentation also applies to recursive procedures written using other types of decision diagrams, in particular, BDDs, ADDs [2], or MTBDDs [7]. Therefore, in this subsection we use the term "DD" instead of "ZDD".

Procedures written with DDs can be roughly divided into two classes:

- Recursive procedures that rely on the DD structure to perform computation.
- Non-recursive procedures that do not use the DD structure but may call the recursive procedures.

The former type is also known as *traversal procedures*, because, in the process of recursion, nodes of the DD are visited in a depth-first manner starting from the root. The efficiency of traversal procedures comes from the fact that, due to the caching of the intermediate results of computation, each node in the tree of recursive calls is visited at most once.

If the traversal procedure takes only one DD as an argument, the number of nodes in the tree is equal to the number of nodes in the DD. If there are more arguments, the number of nodes in the tree is bounded above by the product of the numbers of nodes in the argument DDs. This upper bound is rarely reached in practice.

```
dd TraversalProcedure( dd A, dd B, ... )
{
    // (1) consider trivial cases
    if ( A = 0 )  return ...;
    if ( A = 1 )  return ...;
    ...

    // (2) perform a cache lookup
    R = CacheLookup( A, B, ... );
    if ( R exists )  return R;

    // (3) find the topmost variable in A, B, ...
    Var = TopMostVariable( A, B, ... );

    // (4) cofactor arguments w.r.t. Var
    A0 = Cofactor( A, Var' );
    A1 = Cofactor( A, Var  );
    ...

    // (5) recursively solve subproblems
    R0 = TraversalProcedure( A0, B0, ... );
    R1 = TraversalProcedure( A1, B1, ... );
    ...

    // (6) derive the solution of the problem
    //     from those of the subproblems
    R  = GetResult( R0, R1, ... );

    // (7) cache the result
    CacheInsert( A, B, ..., R );

    // (8) return the result
    return R;
}
```

Figure 1.9: Structure of a recursive DD-based procedure.

Fig. 1.9 shows the structure of a recursive traversal procedure. Steps (2)–(5) and (7)–(8) are similar for the majority of the recursive procedures. Steps (1) and (6) are application-specific.

In particular, step (1) solves the problem in the extreme case when the argument DDs are such that no further recursive calls are necessary. Step (6) answers the question: How does one solve the problem if partial solutions of subproblems are known? In some cases, this step requires much creativity to implement. Some algorithms are not implemented as a single recursive procedure because solutions for Step (6) are not known, for example:

- Complementation of the cover represented by a ZDD. The solution requires two traversals: create the BDD of the function represented by the cover, complement the BDD (this is done in constant time for BDDs with complement edges), and compute the ZDD of the complemented cover.
- Computation of the set of dominated columns and rows in the unate covering problem represented by BDDs. So far, this problem has been solved either explicitly, without DDs, or by applying formulas with quantifiers to the BDDs of dominance relations [20].

1.5   MANIPULATION OF SETS

In this section, we apply the principles of traversal procedures discussed above to the computation of the union of two sets of subsets.

For example, given the two sets of subsets: $A = \{\{a, b\}, \{c\}\}$ and $B = \{\{a, b\}, \{a, c\}\}$, the union of A and B is $\{\{a, b\}, \{a, c\}, \{c\}\}$.

```
set Union( set A, set B )
{
    // (1) consider trivial cases
    if ( A = {} ) return B;
    if ( B = {} ) return A;
    if ( A = B  ) return A;

    // (3) find the topmost variable in A and B
    var x  = TopVariable( A, B );

    // (4) cofactor arguments w.r.t. x
    set A0 = Subset0( A, x );
    set A1 = Subset1( A, x );
    set B0 = Subset0( B, x );
    set B1 = Subset1( B, x );

    // (5) recursively solve subproblems
    set R0 = Union( A0, B0 );
    set R1 = Union( A1, B1 );

    // (6) derive the solution of the problem
    //     from those of the subproblems
    set R  = CreateZdd( x, R1, R0 );

    // (8) return the result
    return R;
}
```

Figure 1.10: Pseudo-code of the union of two sets of subsets.

The pseudo-code of the set-union operator is shown in Fig. 1.10, where steps (2) and (7) (the cache lookup and insert) have been omitted for clarity. Note how the steps (1) and (6) are solved in the pseudo-code.

The trivial cases take place when at least one of the arguments is the empty set (in this case, the union is equal to the other argument) or when the arguments are equal (in this case, the union is any of the sets).

Procedures *Subset0*() and *Subset1*() compute the cofactors of the initial sets. Two cofactor sets are the sets that do not contain the topmost element and the sets that contain the topmost element.

The solution R of the problem can be derived from the solution of the subproblems. To get the subsets with (without) the topmost element, $R1(R0)$, we compute the union of the argument subsets with (without) the topmost element. This is done using two recursive calls to Union().

Next, we create the ZDD with the topmost element x and cofactors $R0$ and $R1$. In this way, we include into the resulting set of subsets all the subsets with (without) the topmost element if they appear with (without) the topmost element in one of the argument sets of subsets, A or B.

Here, the reader is encouraged to do Exercise 1.5.

1.5.1   A CASE STUDY OF THE CUDD SOURCE CODE

Now, consider the source code of the procedure *cuddZddUnion*(), which implements the recursive step of the set-union operator in the CUDD package (Fig. 1.11). The code is taken from file "cuddZddSetop.c" of the CUDD Release 2.3.1. Here, it is reproduced with minor changes to improve its readability.

The procedure *cuddZddUnion*() is called with three arguments: the pointer to the decision diagram manager (*zdd*), and the ZDDs, P and Q. The local variables defined in the function store the levels of the topmost nodes in P and Q (p_top and q_top), the partial results (t and e), and the final result (*res*). Variables t, e, and *res* represent $R0$, $R1$, and R in Fig. 1.10, respectively.

Function *statLine*(), implemented as a macro, is called with the pointer to the manager. It collects statistics about the number of recursive calls. It does not influence the functionality of *cuddZddUnion*().

The next three lines of the code implement the trivial cases. If one of the arguments, P or Q, is the constant-0 zdd node, DD_ZERO(zdd), representing the empty set, the other argument is returned. If the arguments are equal, the union is equal to any of them.

Function *cuddCacheLookup2Zdd*() performs the cache lookup for a recursive procedure with two DD arguments. This function takes four arguments: the pointer to the manager, the pointer to the calling function (used as a signature to distinguish this cache entry from entries created by other functions possibly called with the same argument DDs), and two arguments, P and Q. This function returns NULL if there is no matching entry in the cache; otherwise, it returns the pointer to the ZDD of the result.

Each DD node F in the CUDD package is annotated with the variable number ($F \rightarrow$ *index*). The constant nodes have the variable number equal to CUDD_CONST_INDEX. Because dynamic variable reordering may be periodically performed in the manager, the variable number alone is not enough to find the position of the given DD node in the variable order. To find the position, the levels of nodes should be determined.

The next two if-statements in the source code determine levels, p_top and q_top, of the topmost nodes in the argument DDs, P and Q. The level is the same as the variable index

(CUDD_CONST_INDEX), if the DD node represents the constant function; otherwise it is determined using the mapping of variables into the corresponding levels, which is stored in the array zdd->permZ.

Next, three cases are considered:

- P is higher in the variable order than Q.
- Q is higher in the variable order than P.
- P is on the same level as Q.

We discuss only the last one, the other two being similar. Macros[3] *cuddT*() and *cuddE*() return the "then" and "else" children (cofactors) of the given DD node. In ZDDs, these cofactors correspond to the sets of subsets, in which the topmost element is present (the "then" cofactor) or absent (the "else" cofactor). The cofactors are used in the recursive calls to cuddZddUnion(), which determine the two components of the result: the set of subsets with the topmost element (t) and without it (e).

All the DDs returned by the function calls are checked to determine if they are NULL. If the returned pointer to the node is not NULL, the node is referenced. If the returned pointer is NULL, it means that during the call either (1) the operating system has run out of memory when the CUDD package attempted to extend the node table, or (2) the dynamic variable reordering has been triggered. In both cases, the recursive traversal is interrupted and NULL is returned to the caller. Note also that the intermediate results referenced at some point during computation are dereferenced by calling *Cudd_RecursiveDerefZdd*().

A few remarks should be made regarding the reference-counting conventions employed in the CUDD package. A detailed treatment is given in CUDD User Manual [39].

Decision diagrams are stored in the DD manager as a shared directed acyclic graph of nodes. To mark the nodes that are in use, they are reference-counted. The reference counter of a DD node tells how many times this node participates in the DDs currently present in the manager. Therefore, each time a new DD node is created, its reference counter is incremented by the call to *cuddRef*().

Similarly, each time an old DD node is deleted, its reference counter is decremented. If after decrementing, the reference counter becomes zero, the node is considered "dead" and the reference counters of the successor nodes are decremented in turn. This recursive decrementing of the reference counters is performed by the call to *Cudd_RecursiveDerefZdd*().

Functions in the CUDD package (with very few exceptions mentioned in the user manual) return *non-referenced* DD nodes. It is the responsibility of a calling function to reference the returned node after checking that it is not NULL.

This principle is used in the source code in Fig. 1.11 several times. First, the returned nodes are referenced after each function call, except for the last call to *cuddZddGetNode*(). In this case, there is no need to reference the node to be returned (*res*) because the function should return the

---

[3]A macro is inlined directly into the code, so there is no function call overhead.

```
DdNode *
cuddZddUnion(
  DdManager * zdd,
  DdNode *    P,
  DdNode *    Q)
{
  int p_top, q_top;
  DdNode *t, *e, *res;

  statLine(zdd);

  if ( P == DD_ZERO(zdd) )   return ( Q );
  if ( Q == DD_ZERO(zdd) )   return ( P );
  if ( P == Q )              return ( P );

  /* check cache */
  res = cuddCacheLookup2Zdd(zdd,cuddZddUnion,P,Q);
  if ( res != NULL )  return ( res );

  if ( cuddIsConstant( P ) )
    p_top = P->index;
  else
    p_top = zdd->permZ[P->index];

  if ( cuddIsConstant( Q ) )
    q_top = Q->index;
  else
    q_top = zdd->permZ[Q->index];

  if ( p_top < q_top )
  {
    e = cuddZddUnion( zdd, cuddE( P ), Q );
    if ( e == NULL )  return ( NULL );
    cuddRef( e );
    res = cuddZddGetNode(zdd,P->index,cuddT(P),e);
    if ( res == NULL )
    {
      Cudd_RecursiveDerefZdd( zdd, e );
      return ( NULL );
    }
    cuddDeref( e );
  }
  else if ( p_top > q_top )
  {
    e = cuddZddUnion( zdd, P, cuddE( Q ) );
    if ( e == NULL ) return ( NULL );
    cuddRef( e );
    res = cuddZddGetNode(zdd,Q->index,cuddT(Q),e);
    if ( res == NULL )
    {
      Cudd_RecursiveDerefZdd( zdd, e );
      return ( NULL );
    }
    cuddDeref( e );
  }
  else
  {
    t = cuddZddUnion( zdd, cuddT(P), cuddT(Q) );
    if ( t == NULL )   return ( NULL );
    cuddRef( t );
    e = cuddZddUnion( zdd, cuddE(P), cuddE(Q) );
    if ( e == NULL )
    {
      Cudd_RecursiveDerefZdd( zdd, t );
      return ( NULL );
    }
    cuddRef( e );
    res = cuddZddGetNode( zdd, P->index, t, e );
    if ( res == NULL )
    {
      Cudd_RecursiveDerefZdd( zdd, t );
      Cudd_RecursiveDerefZdd( zdd, e );
      return ( NULL );
    }
    cuddDeref( t );
    cuddDeref( e );
  }
  cuddCacheInsert2( zdd,cuddZddUnion,P,Q,res );
  return ( res );
} /* end of cuddZddUnion */
```

Figure 1.11: CUDD source code implementing the union of two sets represented as ZDD.

non-referenced node. Of course, it would not be an error to reference it by *cuddRef*(*res*) and then dereference it right before returning by *cuddDeref*(*res*). In the CUDD source code, these two steps are skipped for the sake of efficiency.

The last comment is about function *cuddZddGetNode*(). It returns the new ZDD node of the manager (*zdd*) with the given variable (*P → index*) and cofactors (*t* and *e*). Note the order of cofactors in the argument list: first the "then" cofactor, next the "else" cofactor. After the call to *cuddZddGetNode*(), the cofactor DDs (*t* and *e*) should be dereferenced because *cuddZddGetNode*() references them when it creates the new node (*res*). Because cofactor ZDDs (*t* and *e*) are now part of the result *ZDD*(*res*), there is no need for recursive dereferencing by the call to the function *Cudd_RecursiveDerefZdd*(). The cofactor ZDDs can be efficiently dereferenced using *cuddDeref*().

The commutativity of set-union allows us to improve the implementation by ordering the argument ZDDs (P and Q), which tends to increase the hit-rate of the cache.

There are several ways to implement this improvement. One is based on the assumption that the ordering of arguments is given by the ordering of the pointers to the argument DdNode-structures. In this case, it is enough to replace each call to *cuddZddUnion(zdd, A, B)* by the lines.

```
if ( (unsigned)A < (unsigned)B )
    cuddZddUnion( zdd, A, B );
else
    cuddZddUnion( zdd, B, A );
```

Our experiments have shown that this improvement leads to a 5% speedup in applications performing many calls to *Cudd_zddUnion()*, which calls procedure *cuddZddUnion()*.

1.6   MANIPULATION OF CUBE COVERS

This section gives an illustrative example of a traversal procedure dealing with cube covers represented by ZDDs. Fig. 1.12 shows the pseudo-code of the function implementing the product of two cube covers. As in the case of *Union()*, the cache lookups are omitted.

Consider the trivial cases. If any of the covers contains no cubes, the function represented by the cover is the constant-0 function and the product is 0, so the empty cover is returned. If any of the covers is the tautology cube, the product is equal to the other cover. Finally, if the covers are the same, the product is equal to any of them.

```
cover Product( cover A, cover B )                    cover TA1B1 = Product( A1, B1 );
{                                                    cover TA1B2 = Product( A1, B2 );
    if ( A = {} || B = {} ) return {};               cover TA2B0 = Product( A2, B0 );
    if ( A = {{}} ) return B;                        cover TA2B1 = Product( A2, B1 );
    if ( B = {{}} ) return A;                        cover TA2B2 = Product( A2, B2 );
    if ( A = B    ) return A;
                                                     cover R0 = Union( Union(TA0B0, TA0B2), TA2B0 );
    var x  = TopVariable( A, B );                    cover R1 = Union( Union(TA1B1, TA1B2), TA2B1 );
                                                     cover R2 = TA2B2;
    cover A0, A1, A2, B0, B1, B2;                    cover R  = ComposeCover( x, R0, R1, R2 );
    ( A0, A1, A2 ) = DecomposeCover( A, x );         return R;
    ( B0, B1, B2 ) = DecomposeCover( B, x );    }

    cover TA0B0 = Product( A0, B0 );
    cover TA0B2 = Product( A0, B2 );
```

Figure 1.12: Pseudo-code of the product of two covers.

Next, the topmost variable of the cover x is determined. Each variable of the cover is represented by two ZDD variables. This is exploited by the function *TopVariable()*, which is more complex than the function with the same name used in the procedure *Union()*. Next, both argument covers are cofactored into three subcovers containing products with the negative literal (A0 and $B0$), products with the positive literal ($A1$ and $B1$), and products without the given variable ($A2$ and $B2$).

The main part of computation of the cover-product (step (6)) is based on the following equality:

$$
\begin{aligned}
A * B &= (\bar{x}A0 \cup xA1 \cup A2) * (\bar{x}B0 \cup xB1 \cup B2) \\
&= \bar{x}(A0 * B2 \cup A2 * B0 \cup A0 * B0) \\
&\cup\ x(A1 * B2 \cup A2 * B1 \cup A1 * B1) \cup A2 * B2.
\end{aligned}
$$

This equality reduces the computation of the cover-product to seven recursive calls to cover-product of the cofactor DDs. Out of nine possible combinations (each of the three subcovers of A with each of the three subcovers of B), there is no need to consider two combinations, $A0 * B1$ and $A1 * B0$, because the product of the negative literal and the positive literal of the same variable is zero.

Finally, after five two-argument set-union operations which compute three subcovers, $R0$, $R1$, and $R2$, the resulting cover R is composed using the topmost variable.

Another implementation is possible that replaces two (out of seven) internal calls to *Product*() by two extra calls to *Union*(). Because the implementation of *Union*() is simpler, this implementation is more efficient. The alternative implementation is based on the following equality:

$$
\begin{aligned}
A * B &= (\bar{x}A0 \cup xA1 \cup A2) * (\bar{x}B0 \cup xB1 \cup B2) \\
&= \bar{x}(A0 * B2 \cup B0 * (A0 \cup A2)) \\
&\cup\ x(A1 * B2 \cup B1 * (A1 \cup A2)) \cup A2 * B2.
\end{aligned}
$$

Here, the reader is encouraged to do Exercises 1.6, 1.7, and 1.8.

1.7   MIXED ZDD/BDD APPLICATIONS

In this section, we discuss two procedures, which play an important role in the SOP minimization. These are (1) the computation of all primes of the CSF and (2) the computation of an irredundant SOP of the ISF. In both cases, the arguments are represented as BDDs, while the return values are represented by ZDDs. In general, decision-diagram-based procedures may take ZDDs and return BDDs or have other argument assignments.

1.7.1   COMPUTATION OF THE SET OF ALL PRIMES

The recursive approach to the prime computation has been proposed in [32] and implemented using BDDs/ZDDs in [10]. The pseudo-code is shown in Fig. 1.13.

The trivial cases are simple. If the input function is constant-0, the set of primes is empty. If the input function is constant-1, the prime set composed of the tautology cube is returned. Note that the set of subsets that includes only the empty subset, {{}}, represents the cube with no literals, that is, the tautology cube.

```
cover Primes( func F )                    cover P2 = Primes( F0 & F1 );
{                                         cover P0 = Primes( F0 );
    if ( F = 0 ) return {};               cover P1 = Primes( F1 );
    if ( F = 1 ) return {{}};             P0 = P0 # P2;
                                          P1 = P1 # P2;
    var x = TopVariable( F );
                                          cover P = ComposeCover( x, P0, P1, P2 );
    func F0, F1;                          return P;
    (F0, F1) = DecomposeBdd( F, x );    }
```

Figure 1.13: Pseudo-code of the prime set computation.

If it is not a trivial case, the topmost variable in the BDD of F is determined and the function is decomposed with respect to this variable. Next, the problem is solved in three steps.

First, the set of primes ($P2$) belonging to the intersection of cofactors is computed. These primes do not have the topmost variable as the positive or the negative literal.

Second, the set of all primes of the negative cofactor of the function ($P0$) is computed. These primes will have the topmost variable in the negative polarity. Before the topmost literal is added (when composing the result at the end of the procedure), some of them may be identical to the primes in $P2$. After adding the literal corresponding to the topmost variable, some of the cubes in $P2$ will contain the corresponding cubes in $P0$, because cubes in $P0$ have the negative literal while the cubes in $P2$ do not have the literal associated with this variable. Because the contained cubes are, by definition, not primes, they should be removed. This is done by the set-difference operator applied to $P0$ and $P2$.

Similarly, in the third step, we compute the set of all primes with the positive literal ($P1$). Finally, the resulting set of primes is composed from the three subsets, $P0$, $P1$, and $P2$ and returned.

Again, the prime computation can be improved by detecting situations when the given function is unate in its topmost variable. In this case, there is no need to make one out of the three recursive calls to *Primes*() because the primes of the unate function do not have the given variable in any polarity or have it in only one polarity, either negative or positive, depending on the type of unateness.

The mixed BDD/ZDD implementation of *Primes*() is very efficient because it takes a fraction of a second to compute the primes of any function from the Espresso PLA benchmark set, including the so-called hard benchmarks, on a modern computer. To this end, multi-output functions are converted into single-output ones, as shown in [5].

1.7.2 COMPUTATION OF AN IRREDUNDANT SOP

The algorithm for recursive computation of an *ISOP* (irredundant sum-of-products expression) has been proposed in [33]. It has been implemented using decision diagrams in [9][21]. The pseudo-code is given in Fig. 1.14. Symbols "+", "&", and "#" in the pseudo-code represent the Boolean operations OR, AND, and SHARP where $SHARP(a, b) = a\&\bar{b}$.

The procedure *IrrSOP*(), which computes the irredundant sum-of-products, is called with two arguments representing an ISF. The first argument F is the on-set, while the second argument FD is the sum of the on-set and the dc-set. If the ISF is represented by the on-set and off-set, FD is computed by complementing the off-set.

The following are the trivial cases. If the on-set F is empty, the function is a constant-0 function, and the empty cover is returned. If the union of the on-set and dc-set, FD, covers the whole Boolean space, the tautology cube (the set of subsets composed of the empty subset) is returned.

Next, the topmost variable x and the cofactors of F and FD with respect to the variable x are derived. To find the decomposition of BDDs, the ordinary BDD cofactoring with respect to the topmost variable is used. Note that if one of the functions, F or FD, does not depend on the topmost variable, its cofactors with respect to the topmost variable are equal to the function itself.

The ISOP is computed in three steps. The first step finds the ISOP of that part of the on-set F, which cannot be covered by the cubes without the given variable. To achieve this, the ISOP of the part of $F0$ that is outside $FD0$ is computed. This part can be covered only by the cubes that contain the topmost variable in the negative polarity. These cubes are assigned to $R0$.

Similarly, in the second step, the ISOP of the part of the on-set covered only by the cubes with the topmost variable in the positive polarity is computed. These cubes are assigned to $R1$.

In the third and final step, the area in the intersection of $FD0$ and $FD1$, which (1) belongs to F0 or F1 and (2) is not covered by the cubes in $R0$ and $R1$, is computed. The pseudo-code employs the function $Bdd()$, which takes the cover as a ZDD and returns the BDD of the area of the Boolean space corresponding to the cover.

The CUDD package provides the implementation of the ISOP procedure, in which two values are returned in each call: the ZDD of the ISOP cover and the BDD of the area covered by the cover. In this way, there is no need for the call to $Bdd()$. The penalty for this solution is the necessity to cache two values, the ZDD and the BDD, and the repeated recursive call if at least one of the values is lost in the cache (the loss of values in the cache happens when two different computed results hash into the same cache entry resulting in the loss of the earlier result).

The EXTRA library gives an alternative implementation of this procedure, in which only one value is cached and returned. This implementation is based on a specialized operator, which takes two arguments, the BDD of the function and the ZDD of the cover and returns the BDD of the function that is not covered by the cover. This operator can implement expressions of the type $A - Bdd(B)$ in one traversal, while in the pseudo-code of Fig.1.14, expressions of this kind are implemented by two traversals: $Bdd()$ and Boolean SHARP. Experimentation shows that this implementation of an ISOP cover is more efficient than the one proposed in the CUDD package.

```
cover IrrSOP( func F, func FD )              cover R0 = IrrSOP( G0, FD0 );
{                                            func  G1 = F1 - FD0;
   if ( F  = 0 )  return {};                 cover R1 = IrrSOP( G1, FD1 );
   if ( FD = 1 )  return {{}};
                                             func  H  = (F0 - Bdd(R0)) + (F1 - Bdd(R1));
   var x = TopVariable( F, FD );             func  HD = FD0 & FD1;
                                             cover R2 = IrrSOP( H, HD );
   func F0, F1, FD0, FD1;
   (F0, F1) = DecomposeBdd( F, x );          cover R  = ComposeCover( x, R0, R1, R2 );
   (FD0,FD1)= DecomposeBdd( FD, x );         return R;
                                           }
   func  G0 = F0 - FD1;
```

Figure 1.14: Pseudo-code of irredundant SOP computation.

The ISOP computed by this algorithm has remarkable properties [30]. The actual cover depends on the ordering of variables in the DD manager. In some practical cases, the number of cubes in the resulting ISOP is close to that in the exact minimum cover [9][21].

One main advantage of the ZDD-based ISOP is the speed of computing it. For large benchmarks, good-quality covers containing thousands of cubes can be derived in a fraction of a second. Using ISOPs instead of other heuristic algorithms for the SOP computation may lead to a substantial speedup in some applications [19].

Our experiments have also shown that the resulting ISOPs often lead to factored forms with fewer literals than those computed from SOPs minimized using other methods. This may have to do with the fact that the ZDD-based computation enforces a fixed variable order while processing different cofactors of the original function.

1.8   A LIST OF PUBLISHED ZDD APPLICATIONS

ZDDs were introduced in 1993 by Minato [22] and studied in [27][29]. Several ZDD packages were implemented [16][17][25][39]. Here is an incomplete list of problems arising in computer science and engineering that have been successfully solved using ZDDs:

- To represent sets in various applications [24][36].
- To represent cubes and essential primes in two-level SOP minimization [10] and factorization [23][28][38].
- To solve the unate covering problem arising in multi-layer planar routing [12].
- To find dichotomy-based constraint encoding [13][15].
- To solve graph optimization problems [14].
- To represent and manipulate regular expressions under length constraint [18].
- To represent and manipulate polynomials with integer coefficients [26].
- To minimize exclusive SOP [34][35].
- To perform symbolic traversal of FSMs and Petri nets [41][43].
- To perform Davis-Putman resolution procedure [6].

- To synthesize pass-transistor logic [3].
- To find finding all disjoint-support decompositions of completely specified logic functions [30].
- To find unate decomposition of Boolean functions [19].

1.9 CONCLUSIONS

This chapter introduces the reader to the fascinating world of Zero-Suppressed Binary Decision Diagrams. It shows that sparse sets, that is, objects having substantially more zeros than ones, can be more efficiently represented using ZDDs, compared to the traditional BDDs. An implementation of ZDD-based computation is discussed in detail while illustrating it using the actual source code taken from a popular decision diagram package. An extensive list of ZDD-based procedures is given. Finally, it is shown, by way of example, how to use ZDDs for solving a number of computationally hard problems. After studying this chapter, the reader should be able to implement his or her own applications using ZDDs.

1.10 ACKNOWLEDGEMENTS

The author thanks Gianpiero Cabodi, Olivier Coudert, Timothy Kam, Jørn Lind-Nielsen, Shin-ichi Minato, In-Ho Moon, Thomas Shiple, and Fabio Somenzi whose books, papers, source code, and personal advice helped him learn how to implement and apply decision diagrams.

1.11 APPENDIX A

ZDD procedures in CUDD

This section lists the ZDD procedures implemented in the package, as described in the CUDD user manual.

ZDD CONSTRUCTION

Cudd_ReadZero : Returns a contant-0 ZDD node representing the empty set.

Cudd_ReadOne : Returns a contant-1 ZDD node representing the set composed of the empty subset only.

Cudd_ReadZddOne : Returns the ZDD of the constant-1 function assuming that this function depends on the given number of variables.

Cudd_zddIthVar : Returns the ZDD of the function equal to the elementary variable if this variable exists in the DD manager, or creates a new ZDD variable.

Cudd_zddIte : Computes the ITE of three functions represented by ZDDs.

Cudd_zddChange : Substitutes a variable by its complement in a ZDD.

PORTING

Cudd_zddPortFromBdd : Converts a BDD into a ZDD.

Cudd_zddPortToBdd : Converts a ZDD into a BDD.

Cudd_zddVarsFromBddVars : Creates one or more ZDD variables for each BDD variable.

COFACTORING

Cudd_zddSubset0(Cudd_zddSubset1) : Computes the negative (positive) cofactor of a ZDD with respect to a variable.

SET OPERATORS

Cudd_zddUnion : The union of two sets.

Cudd_zddIntersect : The intersection of two sets.

Cudd_zddDiff : The difference of two sets.

Cudd_zddDiffConst : Inclusion test for sets (P implies Q).

Cudd_zddUnateProduct : The product of two unate covers. Unate covers use one ZDD variable for each BDD variable.

Cudd_zddDivide : The quotient of two unate covers.

COVER MANIPULATION

Cudd_zddProduct : The product of two binate covers. The binate covers use two ZDD variables for each BDD variable.

Cudd_zddWeakDiv : The quotient of two binate covers.

Cudd_zddComplement : The complement of a cover.

Cudd_zddIsop : An irredundant sum of products (ISOP) in ZDD form computed from the BDDs for the ON-set and the ON+DC-set.

Cudd_BddFromZddCover : Returns the BDD of the function represented by a cover.

COUNTING FUNCTIONS

Cudd_zddDagSize : Counts nodes in a ZDD.

Cudd_zddCount : Returns the number of paths in a ZDD.

Cudd_zddCountMinterm(Cudd_zddCountDouble) : Count the number of minterms of a ZDD.

REORDERING

Cudd_zddReduceHeap : The main dynamic reordering routine for ZDDs.

Cudd_zddShuffleHeap : Reorders ZDD variables according to given permutation.

Cudd_zddSymmProfile : Prints statistics on symmetric ZDD variables.

REALIGNMENT OF VARIABLES

Cudd_zddRealignEnable : Enables realignment of the ZDD variable order to the BDD variable order after the BDDs and ADDs have been reordered.

Cudd_zddRealignDisable : Disables realignment of ZDD order to BDD order.

Cudd_zddRealignmentEnabled : Returns 1 if the realignment of ZDD order to BDD order is enabled.

PRINTING AND VISUALIZATION

Cudd_zddDumpDot : Writes a file representing the argument ZDDs in a format suitable for the graph drawing program DOT [1].

Cudd_zddPrintCover : Prints an SOP representation of a ZDD.

Cudd_zddPrintDebug : Prints a DD and its statistics to the standard output.

Cudd_zddPrintSubtable : Prints the ZDD table for debugging purposes.

1.12 APPENDIX B

ZDD Procedures in the EXTRA library

The ZDD operators in the library are grouped according to their purpose. Unless stated otherwise, all the operators are implemented in the bottom-up reorder-independent fashion, meaning that an operator restarts when the dynamic variable reordering has taken place.

ZDD CONSTRUCTION

Extra_zddCombination : Creates a ZDD of one combination.

Extra_zddUniverse : Builds a ZDD for all possible combinations of variables from the given set.

Extra_zddTuples : Builds a ZDD representing the set of all tuples of the given cardinality composed of variables from the given set.

Extra_zddRandomSet : Builds the random set of k combinations, each of which may contain up to n elements with the average density d.

Extra_zddConvertBddCubeIntoZddCube : Takes a BDD of the variable product and returns a ZDD cube composed of the same variables.

SET OPERATORS

Extra_zddMaximal : Computes the maximal of the set of elements defined as follows:

$$\max(S) = \{s \in S \mid \forall s' \in S, s \subseteq s' \Rightarrow s = s'\}.$$

Extra_zddMinimal : Computes the minimal of the set of elements defined as follows:

$$\min(S) = \{s \in S \mid \forall s' \in S, s \supseteq s' \Rightarrow s = s'\}.$$

Extra_zddMaxUnion(Extra_zddMinUnion) : Computes the maximal (minimal) of the union of sets X and Y in one bottom-up traversal.

Extra_zddDotProduct : Computes the set of elements created by taking pair-wise unions of elements from X and Y:

$$DotProduct(X, Y) = \{x \cup y | x \in X, y \in Y\}.$$

Extra_zddMaxDotProduct : Computes the maximal of the set creates by taking pair-wise unions of elements from X and Y:

$$MaxDotProduct(X, Y) = \max(\{x \cup y | x \in X, y \in Y\}).$$

Extra_zddSubSet : Computes the set of elements in X that are contained in at least one element of Y:

$$SubSet(X, Y) = \{x \in X | \exists y \in Y, x \subseteq y\}.$$

Extra_zddSupSet : Computes the set of elements in X that contain at least one element of Y:

$$SupSet(X, Y) = \{x \in X | \exists y \in Y, x \supseteq y\}.$$

Extra_zddNotSubSet : Computes the set of elements in X that are not contained in any of the elements of Y:

$$NotSubSet(X, Y) = \{x \in X | \forall y \in Y, x \nsubseteq y\}.$$

Extra_zddNotSupSet : Computes the set of elements in X that do not contain any of the elements of Y:

$$NotSupSet(X, Y) = \{x \in X | \forall y \in Y, x \nsupseteq y\}.$$

Extra_zddMaxNotSupSet : Computes the maximal of the set of elements in X that do not contain any subset of Y in one bottom-up traversal.

Extra_zddEmptyBelongs : Returns 1 if the given ZDD contains the empty combination, and 0 otherwise.

Extra_zddExistAbstract : Removes from a ZDD the occurrences of variables belonging to the given set.

Extra_zddChangeVars : Changes the values of the variables belonging to the given set in all combinations of the ZDD.

Extra_zddCofactor0(Extra_zddCofactor1) : Computes all combinations that contain (do not contain) the variables belonging to the given set.

Extra_zddMaximum(Extra_zddMinimum) : Returns a ZDD representing all combinations of the set S containing the maximum (minimum) number of elements.

Extra_zddSinglesToComb : Takes a ZDD of singleton combinations (combinations including exactly one element) and returns a ZDD containing one combination composed of all elements.

COVER MANIPULATION

Extra_zddPrimes : Given the BDD of the function F, computes a ZDD representing the set of all prime implicants of F.

Extra_zddProductAlt : An alternative implementation of the product of two covers.

Extra_zddPrimeProduct : Computes the product of two covers from which the contained cubes are removed "on the fly."

Extra_zddResolve : Computes all resolvents of the set of clauses S with respect to the set of variables Vars.

Extra_zddCompatible : Computes all the products from the given set that overlap with the given cube.

Extra_zddDisjointCover : Computes the ZDD of the cover represented by disjoint variable paths in the given BDD.

Extra_zddSelectOneCube : Returns a randomly selected product from the given cover.

Extra_zddCheckUnateness : Returns 1 if the given cover is (positive or negative) unate in all its variables.

Extra_zddUnionExor : Computes the union of two covers, while removing the products contained in both covers.

Extra_zddSupercubes : Given two sets of products, computes the set of their pair-wise supercubes.

Extra_zddSelectDist1Cubes : Selects products from the given set that have at least one distance-1 product in another set.

Extra_zddEssential : Computes the essential products.

extraDecomposeCover : Find the cofactors of the cover with respect to the top variable, without creating new DD nodes.

extraComposeCover : Composes the cover from the three subcovers using the given variable.

COVER/FUNCTION MANIPULATION

Extra_zddCoveredByArea : Returns the products from the given cover, completely contained in the given function.

Extra_zddOverlappingWithArea : Returns the products from the given cover, overlapping with the given function.

Extra_zddNotCoveredByCover : Returns the products belonging to the given cover, that are not completely covered by another cover.

Extra_zddNotContainedCubesOverArea : Computes the products that do not overlap with the products from the other set over the given function.

Extra_zddConvertToBdd : An alternative implementation of the procedure *Cudd_MakeBddFromZddCover* from the CUDD package.

Extra_zddConvertToBddAndAdd : Computes the Boolean OR of the given function and the function covered by the cover in one traversal.

Extra_zddConvertEsopToBdd : Computes the BDD of the exclusive sum-of-products represented by a ZDD.

Extra_zddSingleCoveredArea : Computes the function of the Boolean space covered by only one product from the cover.

Extra_zddGetMostCoveredArea : Computes minterms covered by the maximum number of products in a ZDD.

IRREDUNDANT SOP COMPUTATION

Extra_zddIsopCover : A wrapper around *Extra_zddIsop* from the CUDD package. This function returns only the ZDD of the cover and does not return its BDD.

Extra_zddIsopCoverAlt : An alternative implementation of the ISOP computation. This function may be more efficient than *Extra_zddIsop* and *Extra_zddIsopCover*.

Extra_zddIsopCoverRandom : Computes an ISOP cover assuming the random permutation of variables.

Extra_zddIsopCoverAllVars : Tries all possible permutations of variables in every subcover and returns the ISOP with the smallest number of products (potentially very slow for more than 10 variables).

Extra_zddIsopCoverUnateVars : Detects unate variables and performs decomposition w.r.t. these variables first (this function is slower but gives smaller covers compared to *Extra_zddIsop* and *Extra_zddIsopCover*.)

GRAPH INPUT/OUTPUT

Extra_zddGraphRead : Reads the file with the non-directed graph in DIMACS formats and creates its representation as a ZDD.

Extra_zddGraphWrite : Writes the non-directed graph represented as a ZDD into a file in DIMACS ASCII format.

Extra_zddGraphDumpDot : Writes a file representing the graph in a format suitable for the graph-drawing program DOT [1].

GRAPH OPERATORS

Extra_zddCliques : Finds the set of all cliques of the graph represented as a ZDD.

Extra_zddMaxCliques : Finds the set of all maximal cliques of the graph represented as a ZDD in one bottom-up traversal.

Extra_zddIncremCliques : Given a ZDD of the graph and a ZDD of all cliques of size k, computes the set of all cliques of size $k + 1$. (Note that the graph representation is the set of all cliques of the size two.)

Extra_zddGraphColoring : Given a ZDD of the graph, finds a heuristic coloring of the graph (not finished).

Extra_zddRandomGraph : Generates a ZDD representing a random graph with n nodes and density d (not finished).

SET COVERING

Extra_zddSolveUCP : Solves the set-covering problem specified as follows: Each element is encoded using a separate ZDD variable. The only argument S is the set of subsets that covers all elements. The set of elements is determined as the support of S (not finished).

Extra_zddSolveCC : Solves the cyclic core specified by the pair of ZDDs representing the set of rows and the set of columns, using a fast greedy heuristic method.

REORDERING

Extra_zddPermute : Given a ZDD and the permutation of variables, creates a ZDD with permuted variables.

COUNTING FUNCTIONS

Extra_zddLitCount : Counts how many times each element occurs in the combinations of the set.

OTHER FUNCTIONS

Extra_zddSupport : Returns a ZDD representing a set of variables, on which the given DD depends.

Extra_zddVerifyCover : Takes the cover and the function interval represented by two BDDs. Returns 1 if the cover belongs to the interval.

1.13   EXERCISES

1.1. Explain the difference between a decision tree, a BDD, and a ZDD representing the same Boolean function. Draw them for function $F = a \vee b$.

1.2. Suppose there are three elements. Draw ZDDs for (a) the empty subset, (b) the set of subsets containing only empty subset, (c) the subset containing all objects (the full subset), (d) the set of subsets containing only the full subset, (e) the set containing all possible subsets.

1.3. Explain the difference between a ZDD representing a function and a ZDD representing a cover of the function. Draw these two ZDDs for $F = \bar{a}b \vee \bar{c}$.

1.4. Each path in a BDD represents one product of the function. The set of all paths in the BDDs represents a cover of the function. What is known about this cover? Why it is not possible to use BDDs to represent an arbitrary cover of the function?

1.5. What is the difference between unate and binate set operations? How are these differences reflected in the implementation of these operations using ZDDs?

1.6. Recall the difference between the dot-product and the cross-product of two sets of subsets. Can they be implemented by the same traversal procedure?

1.7. Write the pseudo-code of the ZDD traversal procedure to compute the number of literals in the cover presented as a ZDD.

1.8. The threshold operation is applied to the set of subsets to remove subsets containing more than N elements. Describe how this procedure can be implemented using ZDDs.

REFERENCES

[1] AT&T Labs Research. Graphviz. http://www.research.att.com/sw/tools/graphviz/. 25, 28

[2] R. I. Bahar, E. A. Frohm, Ch. M. Gaona, G. D. Hachtel, E. Macii, A. Pardo, F. Somenzi, "Algebraic decision diagrams and their applications," *Proc. of ICCAD' 93*. pp. 188-191. DOI: 10.1023/A:1008699807402. 13

[3] V. Bertacco, S. Minato, R. Verplaetse, L. Benini, G. De Micheli, "Decision diagrams and pass transistor logic synthesis," *Technical Report CSL-TR-97-748*. Computer System Laboratory. Stanford University. December 1997. DOI: 10.1002/cta.822. 2, 23

[4] R. E. Bryant, "Graph-based algorithms for boolean function manipulation," *IEEE Trans. on Comp.*, Vol. C-35, No. 8 (August, 1986), pp. 677-691. DOI: 10.1109/TC.1986.1676819. 1, 4

[5] E. Cerny, M.A. Marin, "An approach to unified methodology of combinational switching circuits," *IEEE Trans. on Computers*, C-26, 8 (Aug, 1977), pp. 745-756. DOI: 10.1109/TC.1977.1674912. 7, 20

[6] P. Chatalic, L. Simon, "ZRes: The old Davis-Putnam procedure meets ZBDDs," *Proc. of CADE-17: Conference on Automated Deduction 2000*, pp. 449-454. DOI: 10.1007/10721959_35. 22

[7] E. M. Clarke, K. L. McMillan, X. Zhao, M. Fujita, J. Yang, "Spectral transforms for large Boolean functions with application to technology mapping," *Proc. ACM/IEEE Design Automation Conference* (DAC) (1993) pp. 54-60, June 1993. DOI: 10.1109/DAC.1993.203919. 13

[8] O. Coudert, C. Berthet, J. C. Madre, "Verification of sequential machines using boolean functional vectors," in *Formal VLSI Correctness Verification*, L. J. M. Claesen Editor, North-Holland, pp. 179-196, Nov. 1989. 1

[9] O. Coudert, J. C. Madre, H. Fraisse, H. Touati, "Implicit prime cover computation: An overview," *Proc. of SASIMI '93*, Nara, Japan. 20, 22

[10] O. Coudert, "Two-level logic minimization: An overview," *Integration*. Vol. 17, No. 2, pp. 97-140, October 1994. DOI: 10.1016/0167-9260(94)00007-7. 2, 19, 22

[11] O. Coudert, "Doing two-level logic minimization 100 times faster," *Proc. of Symposium on Discrete Algorithms (SODA)*, San Francisco CA, January 1995. DOI: 10.1145/313651.313674. 1

[12] O. Coudert, C.-J. R. Shi, "Exact multi-layer topological planar routing," *Proc. of IEEE Custom Integrated Circuit Conference '96 (CICC)*, pp. 179-182. DOI: 10.1109/CICC.1996.510538. 22

[13] O. Coudert, C.-J. R. Shi, "Exact dichotomy-based constraint encoding," *Proc. of ICCD '96*, pp. 426-431. DOI: 10.1109/ICCD.1996.563589. 22

[14] O. Coudert, "Solving graph optimization problems with ZBDDs," *Proc. of ED&T '97*, pp. 224-228. DOI: 10.1109/EDTC.1997.582363. 22

[15] O. Coudert, "A New paradigm of dichotomy-based constraint encoding," *Proc. of DATE '98*, pp. 830-834. DOI: 10.1109/DATE.1998.655954. 22

[16] O. Coudert, TiGeR Package. 22

[17] S. Höreth, TUD Package. 22

[18] S. Ishihara, S. Minato, "Manipulation of regular expressions under length constraints using Zero-Suppressed BDDs," *Proc. of ASP-DAC '95*, pp. 391-396. DOI: 10.1109/ASP-DAC.1995.486250. 22

[19] J. Jacob, A. Mishchenko, "Unate decomposition of Boolean functions," *Proc. of IWLS '01*, Lake Tahoe, California. DOI: 10.1145/2483028.2483138. 2, 22, 23

[20] T. Kam, T. Villa, R. Brayton, A. Sangiovanni-Vincentelli, *Synthesis of Finite State Machines: Functional Optimization*. Kluwer Academic Publishers, 1997. 14

[21] S. Minato, "Fast generation of irredundant sum-of-products forms from binary decision diagrams," *Proc. of SASIMI '92 (Synthesis and Simulation Meeting and International Interchange)*, Kobe, Japan, pp. 64-73. 20, 22

[22] S. Minato, "Zero-suppressed BDDs for set manipulation in combinatorial problems," *Proc. of DAC '93*, pp. 272-277. DOI: 10.1145/157485.164890. 2, 22

[23] S. Minato, "Fast weak-division method for implicit cube representation," *Proc. SASIMI '93*, pp. 423-432, Oct.1993. DOI: 10.1109/43.494701. 22

[24] S. Minato, "Calculation of unate cube set algebra using zero-suppressed BDDs," *Proc. of DAC '94*, pp. 420-424. DOI: 10.1145/196244.196446. 11, 12, 22

[25] S. Minato, ZDD package referenced in [23]. 22

[26] S. Minato, "Implicit manipulation of polynomials using zero-suppressed BDDs," *Proc. of ED&TC '95*, pp. 449-454. DOI: 10.1109/EDTC.1995.470321. 22

[27] S. Minato, *Binary Decision Diagrams and Applications for VLSI CAD*, Kluwer 1995. 11, 22

[28] S. Minato, "Fast factorization method for implicit cube cover representation," *IEEE Trans. CAD*, Vol. 15, No 4, April 1996, pp. 377-384. DOI: 10.1109/43.494701. 2, 11, 12, 22

[29] S. Minato, "Graph-based representations of discrete functions," In T. Sasao, ed., *Representation of Discrete Functions*, Ch. 1, pp. 1-27, Kluwer 1996. DOI: 10.1007/978-1-4613-1385-4. 22

[30] S. Minato, G. DeMicheli, "Finding all simple disjunctive decompositions using irredundant sum-of-products forms," *Proc. of ICCAD '98*, pp. 111-117. DOI: 10.1145/288548.288586. 22, 23

[31] A. Mishchenko, "An introduction to zero-suppressed binary decision diagrams,", *Technical Report*, Portland State University, June 2001. 1

[32] A. Mishchenko, EXTRA Library of DD procedures,
http://www.eecs.berkeley.edu/~alanmi/research/extra 3, 19, 90

[33] E. Morreale, "Recursive operators for prime implicant and irredundant normal form determination," *IEEE Trans. Comp.*, C-19(6), 1970, pp. 504-509. DOI: 10.1109/T-C.1970.222967. 20

[34] H. Ochi, "An exact minimization of AND-EXOR expressions using encoded MRCF," *IEICE Trans. Fundamentals*, Vol. E79-A, No. 12, Dec. 1996, pp. 2131-2133. 22

[35] H. Ochi, "A zero-suppressed BDD package with pruning and its applications to GRM minimization," *IEICE Trans. Fundamentals*, Vol. E79-A, No. 12, Dec. 1996, pp. 2134-2139. 22

[36] H. G. Okuno, S. Minato, H. Isozaki, "On the properties of combination set operations," *Information Processing Letters*, 66 (1998), pp. 195-199. DOI: 10.1016/S0020-0190(98)00067-2. 22

[37] T. Sasao, *Switching Theory for Logic Synthesis*, Kluwer Academic Publishers, 1999. DOI: 10.1007/978-1-4615-5139-3. 1

[38] H. Sawada, S. Yamashita, A. Nagoya, "An efficient method for generating kernels on implicit cube set representations," *Proc. of IWLS '99*, pp. 260-263. 2, 22

[39] F. Somenzi, CUDD Package. `http://vlsi.Colorado.EDU/~fabio/CUDD/cuddIntro.html` 1, 3, 16, 22

[40] G. Swamy, R. Brayton, and P. McGeer, "A fully implicit Quine-McCluskey procedure using BDD's," *Tech. Report No. UCB/ERL M92/127*, 1992. 1

[41] M. Tomisaka, T. Yoneda, "Partial order reduction in symbolic state space traversal using ZBDDs," *IEICE Trans. Fundamentals*, Vol. E00-D, No. 1, Jan.1999, pp. 1-8. 22

[42] T. Villa, T. Kam, R. Brayton, A. Sangiovanni-Vincentelli, *Synthesis of Finite State Machines: Logic Optimization*, Kluwer Academic Publishers, 1997. Chapter 10. "Implicit formulation of unate covering". pp. 301-321. DOI: 10.1007/978-1-4615-6155-2.

[43] T. Yoneda, H. Hatori, A. Takahara, S. Minato, "BDDs vs. Zero-Suppressed BDDs: for CTL symbolic model checking of Petri nets," *Proc. FMCAD '96*. LNCS #1166. pp. 435-449. DOI: 10.1007/BFb0031826. 22

CHAPTER 2

Efficient Generation of Prime Implicants and Irredundant Sum-of-Products Expressions

Tsutomu Sasao

CHAPTER SUMMARY

This chapter shows a method to generate prime implicants and irredundant sum-of-products expressions using the divide and conquer methods [6]. It extends the coverage of this topic presented in Chapter 1.

2.1 LOGICAL EXPRESSIONS

Definition 1.1 *A (two-valued) variable x_i has two* **literals** *x_i and $\bar{x}_i$. A* **product** *is the AND of literals in which no variable occurs more than once. A product is also called a* **product term** *or a* **term**. *A term can be a single literal. A logical sum (OR) of product terms forms a* **sum-of-products expression** *(SOP).*

Example 1.1 *$x_1\bar{x}_2$ and $\bar{x}_1 x_2$ are product terms. $x_1\bar{x}_2 \vee \bar{x}_1 x_2$ is an SOP.* *(End of Example)*

Definition 1.2 *A* **minterm** *is a logical product of all the literals where each variable occurs as exactly one literal. A* **canonical sum-of-products expression** *(canonical SOP) is a logical sum of minterms, where all the minterms are different. It is also called as a* **canonical disjunctive form** *or a* **minterm expansion**.

An arbitrary logic function is represented by its canonical SOP and the representation is unique for a given set of variables. The word **canonical** means that the representation is unique for a logic function. On the other hand, an SOP in which product terms are unrestricted is not canonical; e.g.,in $x_1 x_2 = x_1 x_2 \bar{x}_3 \vee x_1 x_2 x_3$, two SOPs represent the same function.

2.2 MONOTONE AND UNATE FUNCTIONS

A monotone increasing function is a function that can be represented by AND and OR gates only. A unate function is a generalization of a monotone function. Computations of prime implicants and irredundant SOPs (ISOPs) for general logic functions are, in general, quite time-consuming. However, for unate functions, they can be computed very efficiently.

Definition 2.1 *Let **a** and **b** be Boolean vectors. If f satisfies $f(\mathbf{a}) \geq f(\mathbf{b})$, for any vectors such that $\mathbf{a} \geq \mathbf{b}^1$, then f is a* **monotone increasing function**.

Theorem 2.1 *f is a monotone increasing function iff f is a constant or represented by an SOP without complemented literals.*

Example 2.1 *The monotone increasing functions of two variables are: 0, x_1, x_2, $x_1 x_2$, $x_1 \vee x_2$, and 1.*   *(End of Example)*

Definition 2.2 *Let **a** and **b** be Boolean vectors. If f satisfies $f(\mathbf{a}) \leq f(\mathbf{b})$ for all vectors such that $\mathbf{a} \geq \mathbf{b}$, then f is a* **monotone decreasing function**.

Example 2.2 *The monotone decreasing functions of two variables are: 0, $\bar{x}_1$, $\bar{x}_2$, $\bar{x}_1 \bar{x}_2$, $\bar{x}_1 \vee \bar{x}_2$, and 1.*   *(End of Example)*

Theorem 2.2 *f is a monotone decreasing function iff f is a constant or represented by an SOP with complemented literals only.*

A monotone decreasing function can be obtained by complementing a monotone increasing function. Also, a monotone decreasing function can be obtained from a monotone increasing function by complementing all variables.

Definition 2.3 *If a function f is a constant or is represented by an SOP in which every occurrence of a variable is complemented only, or every occurrence of a variable is uncomplemented only, then f is a* **unate function**. *A function which is not unate is a* **binate function**.

The set of unate functions include monotone increasing and monotone decreasing functions as proper subsets.

Example 2.3 *The following functions are represented by SOPs where only either complemented or un-complemented literals appear for each variable. Thus, they are unate functions:*

$$
\begin{aligned}
f_1(x_1, x_2, x_3) &= x_1 \vee x_2 x_3, \\
f_2(x_1, x_2, x_3) &= \bar{x}_1 \bar{x}_2 \vee \bar{x}_3, \\
f_3(x_1, x_2, x_3) &= x_1 \vee x_2 \bar{x}_3.
\end{aligned}
$$

Specifically, f_1 is a monotone increasing function, f_2 is a monotone decreasing function, and f_3 is a unate function. In the following SOP, both x_1 and $\bar{x}_1$ appear. So, one may think it is not a unate function.

$$
f_4(x_1, x_2) = x_1 \vee \bar{x}_1 \bar{x}_2.
$$

[1]Let $\mathbf{a} = (a_1, a_2, \ldots, a_n)$ and $\mathbf{b} = (b_1, b_2, \ldots, b_n)$. Then $\mathbf{a} \geq \mathbf{b}$ iff $(a_i \geq b_i)$ for all $i = 1, 2, \ldots n$.

However, this function is also represented as $f_4(x_1, x_2) = x_1 \vee \bar{x}_2$. Thus, it is a unate function. $f_5(x_1, x_2, x_3) = x_1 x_2 \vee \bar{x}_1 x_3$ is not a unate function. It is a binate function, since any SOP for f_5 contains both $\bar{x}_1$ and x_1.   *(End of Example)*

Here, the reader is encouraged to do Exercise 2.1.

2.3   PRIME IMPLICANTS

In this part, we introduce the concept of the prime implicant, which represents as large as possible loop in a Karnaugh map.

Definition 3.1 *In two logic functions f and g, if $g(x) = 1$ for all x such that $f(x) = 1$, then g **contains** f, denoted by $f \leq g$.*

Definition 3.2 *If a logic function f contains a product c, then c is an **implicant** of f.*

Definition 3.3 *A product P is a **sub-product** of Q, if all the literals in P also appear in Q.*

Example 3.1 *Let $c_1 = x_1 x_2$ and $c_2 = x_1$. Since all the literals (x_1) in the product c_2 also appear in the product c_1, c_2 is a sub-product of c_1.*   *(End of Example)*

Definition 3.4 *Let P be an implicant of a logic function f. If no other implicant Q of f is a subproduct of P, then P is a **prime implicant** (PI) of f.*

Theorem 3.1 *Let f be a unate function of n variables. Then, all the PIs for f can be obtained from any SOP for f by deleting redundant literals.*

2.4   GENERATION OF ALL THE PRIME IMPLICANTS

Theorem 4.1 *[3] [5] Let $PI(f)$ be the set of all the PIs for f. Then, the following holds:*

$$PI(f) \subseteq \bar{x}_1 PI(f_0) \cup x_1 PI(f_1) \cup PI(f_0 \cdot f_1), \tag{2.1}$$

where

$$f = \bar{x}_1 f_0 \vee x_1 f_1.$$

$x_1^* PI(f_1)$ denotes the set of products of $PI(f_1)$ each ANDed with the literal x_1^*, where x_1^* denotes either $\bar{x}_1$ or x_1.

Example 4.1 *Generate all the PIs for the function $g(x_2, x_3, x_4)$ shown in Fig. 2.1.[2] Expand g as*

$$g = \bar{x}_2 g_0 \vee x_2 g_1,$$

---

[2]We use $\{x_2, x_3, x_4\}$ instead of $\{x_1, x_2, x_3\}$, because this example will be reused later in examples in which the variable set $\{x_2, x_3, x_4\}$ is essential.

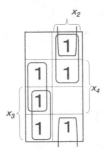

Figure 2.1: Three-variable logic function $g(x_2, x_3, x_4)$ in Example 4.1.

where $g_0 = x_3 \vee x_4$ and $g_1 = \bar{x}_4 \vee \bar{x}_3$. Both g_0 and g_1 are unate functions, thus, PIs for g_0 and g_1 can be obtained by deleting redundant literals and products from their SOPs for g_0 and g_1, respectively. In this case, we have

$$\begin{aligned} PI(g_0) &= \{x_3, x_4\}, \\ PI(g_1) &= \{\bar{x}_3, \bar{x}_4\}. \end{aligned}$$

Next, consider the product function $h = g_0 \cdot g_1$:

$$\begin{aligned} h &= (x_3 \vee x_4)(\bar{x}_4 \vee \bar{x}_3) = \bar{x}_3 x_4 \vee x_3 \bar{x}_4. \\ &= \bar{x}_3 h_0 \vee x_3 h_1, \end{aligned}$$

where $h_0 = x_4$ and $h_1 = \bar{x}_4$. Since they are unate functions, we have

$$\begin{aligned} PI(h_0) &= \{x_4\}, \\ PI(h_1) &= \{\bar{x}_4\}. \end{aligned}$$

Note that $h_0 \cdot h_1 = 0$. From these, we have

$$\begin{aligned} PI(h) &\subseteq \bar{x}_3 PI(h_0) \cup x_3 PI(h_1) \cup PI(h_0 \cdot h_1). \\ &= \{\bar{x}_3 x_4, x_3 \bar{x}_4\}. \end{aligned}$$

Finally, by the relation (2.1) we have

$$\begin{aligned} PI(g) &\subseteq \bar{x}_2 PI(g_0) \cup x_2 PI(g_1) \cup PI(g_0 \cdot g_1) \\ &= \bar{x}_2\{x_3, x_4\} \cup x_2\{\bar{x}_3, \bar{x}_4\} \cup \{\bar{x}_3 x_4, x_3 \bar{x}_4\}. \\ PI(g) &= \{\bar{x}_2 x_3, \bar{x}_2 x_4, x_2 \bar{x}_3, x_2 \bar{x}_4, \bar{x}_3 x_4, x_3 \bar{x}_4\}. \end{aligned}$$

Note that g has 6 PIs.   *(End of Example)*

Next, consider a more complicated function.

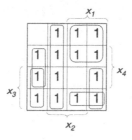

Figure 2.2: Four-variable logic function $f(x_1, x_2, x_3, x_4)$ in Example 4.2.

Example 4.2 *Generate all the PIs for the function $f(x_1, x_2, x_3, x_4)$ shown in Fig. 2.2. First, expand f as*

$$f = \bar{x}_1 f_0 \vee x_1 f_1.$$

Furthermore, expand f_0 and f_1 as

$$f_0 = \bar{x}_2 f_{00} \vee x_2 f_{01},$$
$$f_1 = \bar{x}_3 f_{10} \vee x_3 f_{11}.$$

Fig. 2.3 shows maps for these four subfunctions, where

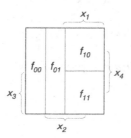

Figure 2.3: Sub-functions of four-variable function in Example 4.2.

$$f_{00} = x_3 \vee x_4,$$
$$f_{01} = 1,$$
$$f_{10} = 1,$$
$$f_{11} = \bar{x}_2 \vee \bar{x}_4.$$

Note that these four functions are unate. Since $f_{00} \subseteq f_{01}$ and $f_{11} \subseteq f_{10}$, f_0 and f_1 are also unate functions. By Theorem 3.1, the PIs for f_0 and f_1 are

$$PI(f_0) = \{x_2, x_3, x_4\},$$
$$PI(f_1) = \{\bar{x}_2, \bar{x}_3, \bar{x}_4\}.$$

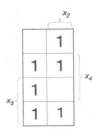

Figure 2.4: Product function $g = f_0 \cdot f_1$ in Example 4.2.

Next, consider the product function $g = f_0 \cdot f_1$, *whose map is shown in Fig. 2.4. Note that the PIs for* g *have been obtained in Example 4.1. Thus, we have*

$$PI(g) = \{\bar{x}_2 x_3, \bar{x}_2 x_4, x_2 \bar{x}_3, x_2 \bar{x}_4, x_3 \bar{x}_4, \bar{x}_3 x_4\}.$$

From the relation (2.1), we have

$$
\begin{aligned}
PI(f) &\subseteq \bar{x}_1 PI(f_0) \cup x_1 PI(f_1) \cup PI(f_0 \cdot f_1) \\
&\subseteq \bar{x}_1 \{x_2, x_3, x_4\} \cup x_1 \{\bar{x}_2, \bar{x}_3, \bar{x}_4\} \cup PI(g) \\
PI(f) &= \{\bar{x}_1 x_2, \bar{x}_1 x_3, \bar{x}_1 x_4, x_1 \bar{x}_2, x_1 \bar{x}_3, x_1 \bar{x}_4, \\
&\quad \bar{x}_2 x_3, \bar{x}_2 x_4, x_2 \bar{x}_3, x_2 \bar{x}_4, x_3 \bar{x}_4, \bar{x}_3 x_4\}.
\end{aligned}
$$

Note that f *has 12 PIs. There are too many PIs to show on a single map. So, these PIs are shown separately in Figs. 2.5, 2.6, and 2.7.*   *(End of Example)*

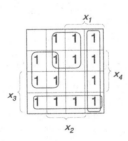

Figure 2.5: PIs for a four-variable function (1) in Example 4.2.

Here, the reader is encouraged to do Exercises 2.2 and 2.3.

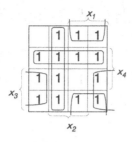

Figure 2.6: PIs for a four-variable function (2) in Example 4.2.

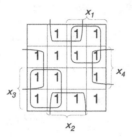

Figure 2.7: PIs for a four-variable function (3) in Example 4.2.

2.5   GENERATION OF IRREDUNDANT SUM-OF-PRODUCTS EXPRESSIONS

Definition 5.1 *Consider an SOP that consists of prime implicants. If any SOP that is obtained by removing any of the prime implicants does not represent the original function f, then the original SOP is an* **irredundant sum-of-products expression** *(ISOP) of f. In an ISOP, each PI covers at least one minterm that is not covered by any other PIs in the ISOP. Among the ISOPs for f, one with the minimum number of products is a* **minimum sum-of-products expression** *(MSOP).*

Theorem 5.1 *Let ISOP(f) be a set of PIs for an ISOP for f. Then, an ISOP(f) can be obtained as*

$$ISOP(f) \subseteq \bar{x}_1 ISOP(f_0) \cup x_1 ISOP(f_1) \cup ISOP(f_0 \cdot f_1)$$

Example 5.1 *Obtain an ISOP for the function g shown in Fig. 2.4. Expand g as $g = \bar{x}_2 g_0 \vee x_2 g_1$, where $g_0 = x_3 \vee x_4$, and $g_1 = \bar{x}_3 \vee \bar{x}_4$. Since both g_0 and g_1 are unate functions, these SOP are also ISOPs for the functions.*

Next, consider the product function

$$h = g_0 \cdot g_1 = (x_3 \vee x_4)(\bar{x}_3 \vee \bar{x}_4) = x_3 \bar{x}_4 \vee \bar{x}_3 x_4.$$

Let $h = \bar{x}_3 h_0 \vee x_3 h_1$, where $h_0 = x_4$ and $h_1 = \bar{x}_4$. From Theorem 5.1, $h = x_3 \bar{x}_4 \vee \bar{x}_3 x_4$ is an ISOP.

Thus, the ISOP for g is obtained as

$$ISOP(g) = \bar{x}_2\{x_3, x_4\} \cup x_2\{\bar{x}_3, \bar{x}_4\} \cup \{x_3 \bar{x}_4, \bar{x}_3 x_4\}.$$

By removing redundant products, we have

$$g = \bar{x}_2 x_3 \vee \bar{x}_2 x_4 \vee x_2 \bar{x}_3 \vee x_2 \bar{x}_4.$$

(End of Example)

Example 5.2 *Obtain an ISOP for the function f shown in Fig. 2.2. First, expand f into:*

$$f = \bar{x}_1 f_0 \vee x_1 f_1,$$

where $f_0 = x_2 \vee x_3 \vee x_4$ and $f_1 = \bar{x}_2 \vee \bar{x}_3 \vee \bar{x}_4$.

By Theorem 5.1, we have

$$ISOP(f) \subseteq \bar{x}_1 ISOP(f_0) \cup x_1 ISOP(f_1) \cup ISOP(f_0 \cdot f_1).$$

Note that both f_0 and f_1 are unate, and ISOPs for them are $f_0 = x_2 \vee x_3 \vee x_4$ and $f_1 = \bar{x}_2 \vee \bar{x}_3 \vee \bar{x}_4$, respectively.

Next, consider the product function $g = f_0 \cdot f_1$, whose map is shown in Fig. 2.4. Note that this is the same function as the one in Example 5.1. So,

$$g = \bar{x}_2 x_3 \vee \bar{x}_2 x_4 \vee x_2 \bar{x}_3 \vee x_2 \bar{x}_4$$

is an ISOP. Thus,

$$\begin{aligned}
ISOP(f) \quad &\subseteq \quad \bar{x}_1\{x_2, x_3, x_4\} \cup x_1\{\bar{x}_2, \bar{x}_3, \bar{x}_4\} \\
&\quad \cup \{\bar{x}_2 x_3, \bar{x}_2 x_4, x_2 \bar{x}_3, x_2 \bar{x}_4\}. \\
&\subseteq \quad \{\bar{x}_1 x_2, \bar{x}_1 x_3, \bar{x}_1 x_4, x_1 \bar{x}_2, x_1 \bar{x}_3, x_1 \bar{x}_4, \\
&\quad \bar{x}_2 x_3, \bar{x}_2 x_4, x_2 \bar{x}_3, x_2 \bar{x}_4\}.
\end{aligned}$$

Note that last four products are redundant, and can be deleted without changing the function represented by the expression. Since no other PI can be deleted, an ISOP for f is

$$f = \bar{x}_1 x_2 \vee \bar{x}_1 x_3 \vee \bar{x}_1 x_4 \vee x_1 \bar{x}_2 \vee x_1 \bar{x}_3 \vee x_1 \bar{x}_4.$$

Fig. 2.8 shows the map of the ISOP. It has six PIs, and has the maximum number of products [7].

(End of Example)

Here, the reader is encouraged to do Exercises 2.4 and 2.5.

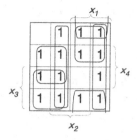

Figure 2.8: ISOP for a four-variable function f in Example 5.2.

2.6   MORREALE'S ALGORITHM

In principle, Theorem 5.1 can be used to generate an ISOP for a given function. However, as shown in Example 5.2, it often generates redundant products. It is time-consuming to remove redundant products. Morreale [3] [3] invented a method to generate only irredundant products:

Algorithm 6.1

1. *Let $f(x)$ be a given incompletely specified logic function. Let $DC(f)$ be the set of all binary vectors x such that the value of $f(x)$ is undefined.*

2. *If there is no vector x such that $f(x) = 1$, then $ISOP(f) \leftarrow 0$. Return.*

3. *If there is no vector x such that $f(x) = 0$, then $ISOP(f) \leftarrow 1$. Return.*

4. *Let x_i be a variable of f.*
   $f_0 \leftarrow f(x|x_i = 0)$
   $f_1 \leftarrow f(x|x_i = 1)$
   $f_0' \leftarrow f_0 \cdot \bar{f_1}, \ DC(f_0') \leftarrow f_0 \cdot f_1$
   $f_1' \leftarrow f_1 \cdot \bar{f_0}, \ DC(f_1') \leftarrow f_0 \cdot f_1$
   $ISOP_0 \leftarrow ISOP(f_0'),$
   / Generate ISOP for f_0', recursively./
   $ISOP_1 \leftarrow ISOP(f_1'),$
   /*Generate ISOP for f_1', recursively.*/
   Let g_0 and g_1 be functions represented by $ISOP(f_0')$ and $ISOP(f_1')$, respectively.
   $f_0'' \leftarrow f_0, \ DC(f_0'') \leftarrow g_0 \cup DC(f_0)$
   $f_1'' \leftarrow f_1, \ DC(f_1'') \leftarrow g_1 \cup DC(f_1)$
   $f_* \leftarrow f_0'' \vee f_1'', \ DC(f_*) \leftarrow DC(f_0'') \cap DC(f_1'')$
   $ISOP_* \leftarrow ISOP(f_*)$
   /* Obtain $ISOP(f_*)$, recursively.*/

---

[3]Since Morreale's paper is hard to read, readers may read Minato's paper [4] instead.

$ISOP(f) \leftarrow \bar{x}_i ISOP_0 \cup x_i ISOP_1 \cup ISOP_*.$

Return.

Example 6.1

1. Consider the four-variable function that appeared in Example 5.2. Fig. 2.9 shows f_0 and f_1. The labels of the variables are same as that of Fig. 2.2.

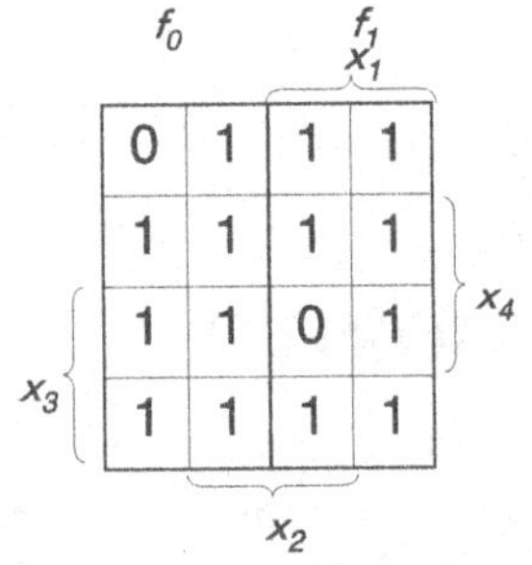

Figure 2.9: $f_0 = f(|x_1 = 0)$ and $f_1 = f(|x_1 = 1)$ in Example 6.1.

2. Fig. 2.10 shows f'_0 and f'_1, $ISOP_0$ and $ISOP_1$. f'_0 denotes the function to be covered by an ISOP including $\bar{x}_i$. f'_1 denotes the function to be covered by an ISOP including x_i.

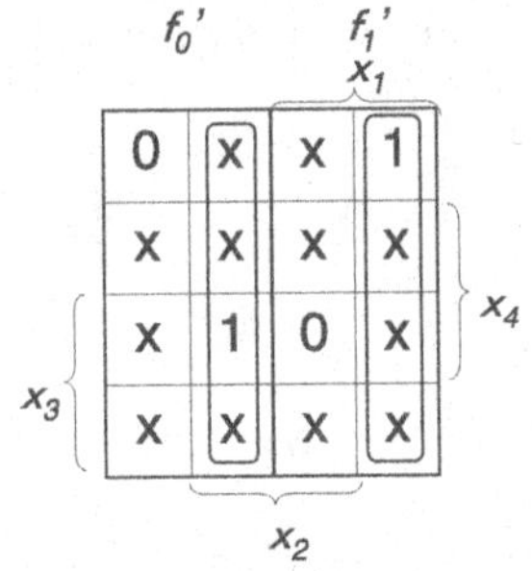

Figure 2.10: $ISOP_0$ and $ISOP_1$ for f'_0 and f'_1 in Example 6.1.

3. Fig. 2.11 shows f''_0 and f''_1. f''_0 denotes the function by assigning don't cares values to the minterms which are already covered by $ISOP_0$. f''_1 denotes the function by assigning don't cares values to the minterms which are already covered by $ISOP_1$.

4. Fig. 2.12 shows f_* and $ISOP_*$. f_* denotes the function to be covered by the ISOP without the literals of x_1.

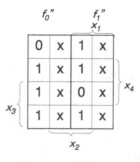

Figure 2.11: f_0'' and f_1'' in Example 6.1.

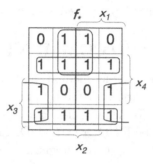

Figure 2.12: $ISOP_*$ in Example 6.1.

5. *Note that*

$$
\begin{aligned}
ISOP_0 &= \bar{x}_1 x_2 \\
ISOP_1 &= x_1 \bar{x}_2, \\
ISOP_* &= x_2 \bar{x}_3 \vee \bar{x}_2 x_3 \vee x_3 \bar{x}_4 \vee \bar{x}_3 x_4.
\end{aligned}
$$

Finally, we have

$$ISOP(f) = \{\bar{x}_1 x_2, x_1 \bar{x}_2, x_2 \bar{x}_3, \bar{x}_2 x_3, x_3 \bar{x}_4, \bar{x}_3 x_4\}.$$

This ISOP has six PIs. Note that Figs. 2.5 and 2.6 show MSOPs for f and, each has only four PIs. (End of Example)

Most functions are binate, so we need Morreale's algorithm to find ISOPs. However, if we know the given function is unate, we have a more efficient method:

Theorem 6.1 *Let f be a unate function of n variables. Then, the unique MSOP is obtained from any SOP for f by deleting redundant literals and products.*

Example 6.2 *Consider the function*

$$f = x_1(x_2 \lor x_3 \lor x_4) \lor x_2(x_3 \lor x_4) \lor x_3 x_4.$$

Since f can be represented without complemented literals, it is a monotone increasing function. Thus, by Theorem 3.1, its PIs can be directly obtained from the SOP for f :

$$f = x_1 x_2 \lor x_1 x_3 \lor x_1 x_4 \lor x_2 x_3 \lor x_2 x_4 \lor x_3 x_4.$$

Also, by Theorem 6.1, it is the unique ISOP, and also the MSOP for this function.   *(End of Example)*

Logic functions used in control circuits are often unate. The algorithm in [1] achieves a reduced computation time by identifying unate variables.

Here, the reader is encouraged to do Exercises 2.6 and 2.7.

2.7   CONCLUSION AND COMMENTS

This chapter showed a divide and conquer method to generate all the prime implicants (PIs) of a given logic function. Although various methods to generate PIs of a logic function have been developed, the presented method is the most efficient.

As for the data structure, both BDDs and ZDDs can be used. In fact, the first program was developed with BDD package [4]. Later, programs using ZDDs were developed.

2.8   EXERCISES

2.1. Consider the following functions:

$$
\begin{aligned}
f_1 &= x_1 x_2 \lor \bar{x}_1 \bar{x}_2 \\
f_2 &= x_1 \lor \bar{x}_1 x_2 \\
f_3 &= x_1 \lor \bar{x}_1 \bar{x}_2 \\
f_4 &= x_1 \lor x_2 x_3 \\
f_5 &= x_1 x_2 \lor \bar{x}_2 x_3
\end{aligned}
$$

1. Identify the monotone increasing functions.

2. Identify the unate functions.

3. Identify the binate functions.

2.2. Derive all the PIs for

$$f = (x_1 \lor x_2 \lor x_3 \lor x_4 \lor x_5)(\bar{x}_1 \lor \bar{x}_2 \lor \bar{x}_3 \lor \bar{x}_4 \lor \bar{x}_5).$$

2.3. How many PIs does the following function has?

$$f_n = (x_1 \vee x_2 \vee \ldots \vee x_n)(\bar{x}_1 \vee \bar{x}_2 \vee \ldots \vee \bar{x}_n).$$

2.4. Derive all the ISOPs for the function $f_1 = x_1\bar{x}_2 \vee x_2\bar{x}_3 \vee x_3\bar{x}_1$.

2.5. Consider the five-variable function:

$$f = x_1\bar{x}_2 \vee x_2\bar{x}_3 \vee x_3\bar{x}_4 \vee x_4\bar{x}_5 \vee x_5\bar{x}_1.$$

1. Derive all the PIs for f using Theorem 4.1.
2. Derive an ISOP for f using Algorithm 6.1.

2.6. Prove the following:

$$(x_1 \vee x_2 \vee \ldots \vee x_n)(\bar{x}_1 \vee \bar{x}_2 \vee \ldots \vee \bar{x}_n) = x_1\bar{x}_2 \vee x_2\bar{x}_3 \vee \ldots \vee x_{n-1}\bar{x}_n \vee x_n\bar{x}_1.$$

2.7. Prove that an n-variable function has at most 3^n PIs.

REFERENCES

[1] R. K. Brayton, G. D. Hachtel, C. T. McMullen, and A. L. Sangiovanni-Vincentelli, *Logic Minimization Algorithms for VLSI Synthesis*, Boston, MA. Kluwer Academic Publishers, 1984. DOI: 10.1007/978-1-4613-2821-6. 46

[2] O. Coudert and J. C. Madre, "Implicit and incremental computation of primes and essential primes of Boolean functions," *Proc. 29th Design Automation Conference*, CA. USA. pp.36-39, June 1992. DOI: 10.1109/DAC.1992.227866.

[3] E. Morreale, "Recursive operators for prime implicant and irredundant normal form determination," *IEEE Transactions on Computers*, Vol. C-19, No. 6, pp. 504-509, June 1970. DOI: 10.1109/T-C.1970.222967. 37, 43

[4] S. Minato, "Fast generation of prime-irredundant covers from binary decision diagrams," *IEICE Trans. Fundamentals*, Vol. E76-A, No. 6, pp. 976-973, June 1993. 43, 46

[5] B. Reusch, "Generation of prime implicants from subfunctions and a unifying approach to the covering problem," *IEEE TC*, Vol. C-24, No. 9, pp. 924-930, Sept. 1975. DOI: 10.1109/T-C.1975.224338. 37

[6] T. Sasao, *Switching Theory for Logic Synthesis*, Kluwer Academic Publishers, 1999. DOI: 10.1007/978-1-4615-5139-3. 35

[7] T. Sasao and J. T. Butler, "Worst and best irredundant sum-of-products expressions," *IEEE Transactions on Computers*, Vol. 50, No. 9, Sept. 2001, pp. 935-948. DOI: 10.1109/12.954508. 42

CHAPTER 3

The Power of Enumeration–BDD/ZDD-Based Algorithms for Tackling Combinatorial Explosion

Shin-ichi Minato

CHAPTER SUMMARY

Discrete structure manipulation is a fundamental technique for many problems solved by computers. BDDs/ZDDs have attracted a great deal of attention for 20 years, because they efficiently manipulate basic discrete structures such as logic functions and sets of combinations.

Recently, one of the most interesting research topics related to BDDs/ZDDs is the "frontier-based method," a very efficient algorithm for enumerating and indexing the subsets of a graph that satisfies a given constraint. This work is important because many kinds of practical problems can be efficiently solved by some variations of this algorithm. In this article, we present an overview of the frontier-based method and recent topics on the state-of-the-art algorithms to show the power of enumeration.

3.1 INTRODUCTION

Discrete structures are foundational material for computer science and mathematics, which are related to set theory, symbolic logic, inductive proof, graph theory, combinatorics, probability theory, etc. Many problems solved by computers can be decomposed into discrete structures using simple primitive algebraic operations. It is very important to represent discrete structures compactly and to execute efficiently tasks, such as equivalency/validity checking, analysis of models, and optimization. Those techniques are commonly used in many application areas in computer science, for example, hardware/software system design, fault analysis of large-scale systems, con-

A preliminary version of this chapter appeared as [9].

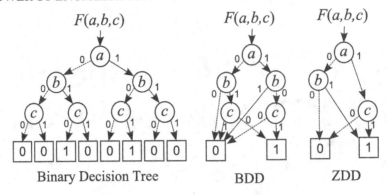

Figure 3.1: Binary decision tree, BDDs and ZDDs.

straint satisfaction problems, data mining, knowledge discovery, machine learning/classification, bioinformatics, and web data analysis.

A binary decision diagram (BDD) is a representation of a Boolean function, one of the most basic models of discrete structures. After the epoch-making paper [1] by Bryant in 1986, BDD-based methods have attracted a great deal of attention. The BDD was originally invented for the efficient Boolean function manipulation required in VLSI logic design, but Boolean functions are also used for modeling many kinds of combinatorial problems. A zero-suppressed decision diagram (ZDD) [8] is a variant of the BDD, customized for manipulating "sets of combinations." ZDDs have been successfully applied not only to VLSI design, but also for solving various combinatorial problems, such as constraint satisfaction, frequent pattern mining, and graph enumeration. Recently, ZDDs have become more widely known, since D. E. Knuth discussed in detail ZDD-based algorithms in the latest volume of his famous series of books [7].

Although a quarter of a century has passed since Bryant first put forth his idea, there are still many interesting and exciting research topics related to BDDs and ZDDs. For example, Knuth presented a surprisingly fast algorithm "Simpath" [7] to construct a ZDD which represents all the paths connecting two points in a given graph structure. This work is important because many kinds of practical problems are efficiently solved by some variations of this algorithm. We generically call such ZDD construction methods "frontier-based methods."

This chapter presents recent research activity related to BDDs and ZDDs. We first briefly explain the basic techniques for BDD/ZDD manipulation. Next, we present an overview of the frontier-based method for efficiently enumerating and indexing the solutions of combinatorial problems. Next, we discuss state-of-the-art algorithms that are effective in the context of combinatorial explosion.

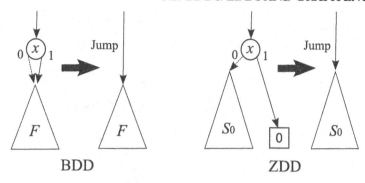

Figure 3.2: BDD and ZDD reduction rules.

3.2 BDDS/ZDDS AND GRAPH ENUMERATION

A binary decision diagram (BDD) is a graph representation for a Boolean function, developed for VLSI design. As illustrated in Fig. 3.1, a BDD is derived by reducing a binary decision tree, which represents a decision making process by the input variables. If we fix the order of input variables, and apply the following two reduction rules, then we have a compact and canonical form [1] for a given Boolean function:

(1) Delete all redundant nodes whose two edges have the same destination (Fig. 3.2), and

(2) Share all equivalent nodes having the same child nodes and the same variable.

The compression ratio achieved by using a BDD compared to a decision tree depends on the property of Boolean function to be represented, but it can be 10 to 100 times in some practical cases. In addition, we can systematically construct a BDD as the result of a binary logic operation (i.e., AND, OR) for a given pair of operand BDDs. This algorithm is based on hash table techniques, and the computation time is almost linear in the BDD size, as measured by the number of nodes.

BDDs can be efficiently manipulated in main memory. Recently, PCs come equipped with gigabytes of main memory, and we can solve larger-scale problems which used to be impossible because of a lack of memory. This has allowed BDDs to be applied to more applications, especially after 2000.

A zero-suppressed decision diagram (ZDD) is a variant of the BDD, customized for manipulating sets of combinations. This data structure was first introduced by Minato [8]. ZDDs are based on the special reduction rules different from those used in BDDs, as follows.

(1) Delete all nodes whose 1-edge directly points to the 0-terminal node. (Fig. 3.2)

This new reduction rule is extremely effective, if it is applied to a set of sparse combinations. If each item appears in 1% of combinations on average, ZDDs are more compact than ordinary

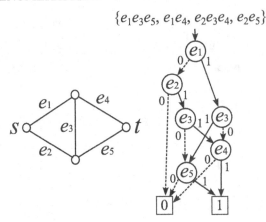

Figure 3.3: ZDD representing the paths from s to t.

BDDs, by up to 100 times. Such situations often appear in real-life problems. For example, in a supermarket, the number of items in a customer's basket is usually many fewer than all the items in the supermarket. Because of such an advantage, ZDDs are now widely recognized as the most important variant of BDDs. Recently, D. Knuth presented a new fascicle of his famous book series [7], which has a section of ZDDs discussed in depth in 30 pages with 70 exercises.

ZDDs can be utilized for enumerating and indexing the solutions of a graph problem. When we assume a graph $G = (V, E)$ with the vertex set $V = \{v_1, v_2, \ldots, v_n\}$ and the edge set $E = \{e_1, e_2, \ldots, e_m\}$, a graph enumeration problem is to compute a subset of the power set 2^E (or 2^V), each element of which satisfies a given property. In this model, we can consider that each solution is a combination of edges (or vertices), and a set of solutions can be represented by a ZDD. For example, Fig. 3.3 shows the ZDD representing the set of paths connecting the two vertices s and t of the graph on the left. Each path can be represented as a combination of the edges, $\{e_1e_3e_5, e_1e_4, e_2e_3e_4, e_2e_5\}$. The ZDD also has four paths from the root node to the 1-terminal node, and each path corresponds to a solution of the problem, where $e_i = 1$ means to use the edge e_i, and $e_i = 0$ means not to use e_i. At this point, the reader may benefit from working Exercise 3.1.

For solving graph enumeration problems, one may ask which data structure is more efficient, a BDD or a ZDD? It depends on the problem to be solved. BDDs and ZDDs are both effective for representing sets of combinations including many similar sub-combinations, because in such cases, their sub-graphs are shared very well. In addition, BDDs are more compact than ZDDs if we often observe that choosing a certain edge makes some other edges don't care. On the other hand, ZDDs are more compact than BDDs if we often observe that choosing a certain edge excludes some other edges. In many cases of graph enumeration problems, ZDDs are likely better than ordinary BDDs. The reader may benefit from working Exercises 3.2 and 3.3.

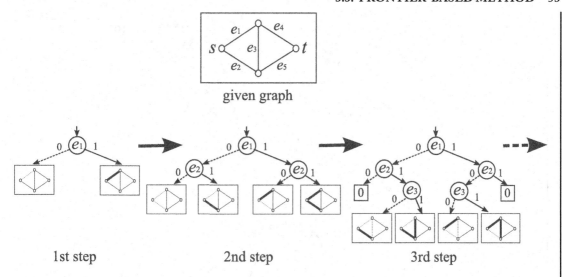

Figure 3.4: Tree expansion in the Simpath algorithm.

3.3   FRONTIER-BASED METHOD

3.3.1   KNUTH'S SIMPATH ALGORITHM

In 2009, Knuth published the surprisingly fast algorithm "Simpath" [7] (Vol. 4, Fascicle 1, p. 121, or p. 254 of Vol. 4A) to construct a ZDD which represents all the simple (or self-avoiding) paths connecting two points s and t in a given graph (not necessarily the shortest ones but ones not passing through the same point twice). This work is important because many kinds of practical problems can be efficiently solved by some variations of this algorithm. Knuth provides his own C source codes on his webpage for public access, and the program is surprisingly fast. For example, in a 14×14 grid graph[1] (420 edges in total), the number of self-avoiding paths between opposite corners is exactly 227449714676812739631826459327989863387613323440 ($\approx 2.27 \times 10^{47}$) ways. By applying the Simpath algorithm, the set of paths can be compressed into a ZDD with only 144759636 nodes, and the computation time for the ZDD generation is only a few minutes. Once we have a compact ZDD, the exact number of the solutions can be counted by a depth-first traversal of all ZDD nodes, called the "Count" operation [8], in linear time with respect to the ZDD size.

Figure 3.4 illustrates the basic mechanism of the Simpath algorithm. In the beginning, we assign a fixed ordering for all the edges, $E = \{e_1, e_2, \ldots, e_n\}$ for the given graph $G = (V, E)$. Then, we construct a binary decision tree from the top to the bottom in a breadth-first manner. In the first step, we consider two decisions 1 and 0, representing whether the edge e_1 is used in

[1]A graph with 14×14 edges, and 15×15 nodes.

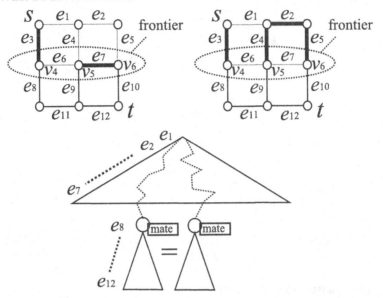

Figure 3.5: Equivalent states and frontiers in the Simpath algorithm.

the s-t path or not. We then make two leaf nodes, each of which holds the current status of the path selection. In the second step, we visit each of the two leaf nodes and expand a new branch from the leaf, to decide whether the edge e_2 is used in the s-t path or not. Then, each of the new leaves has the current status of e_1 and e_2. In this way, at the k-th level, we sequentially visit all the leaf nodes and append a $(k + 1)$-th level decision node for each case, but we may prune the branch when we find a contradiction in the current status such that the edges forming an s-t path cannot avoid generating a disjoint component or a branch. In such cases, we assign the value 0 to the leaf node, and we put no more branches there. By repeating this process until the n-th level, we can construct a total decision tree including all the s-t paths. We assign the value 1 to each final leaf node which represents a solution. After that, we apply the ZDD reduction rules to each node of the decision tree from the bottom to the top to obtain a compressed ZDD. At this point, the reader may benefit from working Exercise 3.5.

In the above procedure, we can avoid an unnecessary expansion by assigning a 0-terminal node to the contradicted node (representing partial pattern which can never be a solution), but this is not enough for really fast computation. The Simpath algorithm also employs another idea of reduction to check equivalent nodes in the k-th level, and those equivalent nodes are merged into one node in the next expansion step. Here, "equivalent nodes" mean that the two intermediate nodes have exactly the same requirements for the remaining undecided edges to complete the correct s-t paths. An example on a grid graph is shown in Fig. 3.5. Suppose that it has already been decided whether e_1 to e_7 are used, and let us compare the two cases where (e_3, e_7) are chosen (left one) and (e_2, e_3, e_4, e_5) are chosen (right one). In both cases, we should connect the vertices v_4 and v_5, and connect the vertices v_6 and t by using remaining edges e_8 to e_{12}. Thus, the requirements of the two cases are completely the same. In the corresponding ZDD, the two sub-graphs starting from e_8 must be identical, so we do not have to traverse such subgraphs twice if we can identify such equivalent nodes. In many cases of the s-t path problem, a number of equivalent nodes appear, and the effect on the computation time is very large.

For checking the equivalence of two leaf nodes, we need only the status on a set of the vertices $\{v_4, v_5, v_6\}$ specified by the dotted circle in the figure, where each vertex connects at least one decided edge and at least one undecided edge. It is enough to know which vertex is at the end point and which one is at the opposite end. The Simpath algorithm prepares an array structure called the "mate" to store that information at each leaf node, and registers all the mate data into a hash table for fast equivalence checking. (For example, $mate[v_4] = S, mate[v_5] = v_6$, and $mate[v_6] = v_5$ in Fig. 3.5.) Knuth called such a set of vertices the "frontier." The frontier area moves from the start vertex to the goal vertex during computation. The reader may benefit from working Exercises 3.6 and 3.7.

The Simpath algorithm is a type of dynamic programming, based on the mate information on the frontier. If the frontier grows larger in the computation process, more intermediate states appear and more computation time is required. Thus, it is important to keep the frontier small. The maximum size of the frontier depends on the given graph structures and the order of the edges. Planar and narrow graphs tend to have small frontiers.

As a related previous work, in 1995, Sekine, Imai et al. [10] proposed a method for generating BDDs to calculate Tutte polynomials [12], representing a kind of graph invariant property. This method is very similar to the Simpath algorithm, but unfortunately, Knuth did not know of their work when publishing his algorithm. The differences are only that they generate BDDs instead of ZDDs, and that the "mate data" represents a set of partitions of connected components, not a pair of end points of paths. The other parts of the algorithm are similar. They also presented extensive mathematical discussions and a theoretical analysis of the complexity of the algorithm for some special classes of graphs, such as planar and grid graphs [3]. Their theoretical results can also be used for the Simpath algorithm in an almost identical manner.

3.3.2 FRONTIER-BASED METHOD FOR VARIOUS PROBLEMS

Knuth states in his book [7] that the Simpath algorithm can easily be modified to generate not only s-t paths but also Hamilton paths, directed paths, some kinds of cycles, and many other problems by slightly changing the mate data structure. We call such ZDD construction methods "frontier-based methods."

The frontier-based methods are different from the conventional ZDD construction. Usually we construct ZDDs by repeating primitive set operations between two ZDDs. In general, those primitive set operations are efficiently implemented based on Bryant's Apply algorithm, but do not directly use the properties of the specific problem. Frontier-based methods are sometimes much more efficient because they are a dynamic programming method based on the frontier, a kind of structural property of the given graph.

Here we list graph problems which can be enumerated and indexed by a ZDD using a frontier-based method.

- all s-t paths,
- s-t paths with length k,
- k-pairs of s-t paths,
- all cycles,
- cycles with length k,
- Hamilton paths / cycles,
- directed paths / cycles,
- all connected components,
- spanning trees / forests,
- Steiner trees,
- all cutsets,
- k-partitioning,
- calculating probability of connectivity
- all cliques,
- all independent sets,
- graph colorings,
- tilings,
- perfect / imperfect matching.

These problems are strongly related to many kinds of real-life problems. For example, path enumeration is of course important in geographic information systems, and is also used for dependency analysis of a process flow chart, fault analysis of industrial systems, etc. Recently, Inoue et al. [4] discussed the design of electric power distribution systems. Such civil engineering systems are usually near to planar graphs, so the frontier-based method is very effective in many cases. They succeeded in generating a ZDD to enumerate all the possible switching patterns in a realistic benchmark of an electric power distribution system with 468 switches. The obtained ZDD represents as many as 10^{60} of the valid switching patterns but the actual ZDD size is less 100 MBytes, and computation time is around 30 minutes. After generating the ZDD, all valid switching patterns are compactly represented, and we can efficiently discover the switching patterns with maximum, minimum, and average cost. Those cost values can be calculated by a depth-first traversal of all ZDD nodes, and the computation time is bounded by the ZDD size. We can also efficiently apply additional constraints to the current solutions. In this way, frontier-based methods can be utilized for many kinds of real-life problems.

Figure 3.6: Screenshots of the animation video [2].

3.3.3   RECENT TOPICS ON THE PATH ENUMERATION PROBLEM

In 2012, the author had a chance to collaborate with "Miraikan" (National Future Science Museum of Japan) to design an exhibition presenting the problem of combinatorial explosion and the state-of-the-art techniques for solving such hard problems. As a work of the exhibition, we supervised a short educational animation video (Fig. 3.6). The video is mainly designed for junior high school to college students, so it does not use any difficult technical terms. It is something like a funny science fiction story.

In this video, we used the simple path enumeration problem for $n \times n$ grid graphs. The story is that the teacher counts the total number of paths for children starting from $n = 1$, but she will be faced with a difficult situation, since the number grows unbelievably fast. She would spend 250,000 years to count the paths for the 10×10 grid graph by using a supercomputer if she used a naïve method. The story ends with a surprise: a state-of-the-art algorithm can finish the same problem in a few seconds.

The video is now shown in the official museum channel of YouTube [2] and surprisingly received 1.4 million views, which is extraordinary in the case of scientific educational contents. In addition, it was our great pleasure to hear that Knuth also enjoyed this video and shared it with several friends.

The video story tells the computation result up to $n = 11$, but it is interesting to know the largest n for which this problem is still computable. We have worked for improving the algorithm for solving the large-scale problems as much as possible, and succeeded in counting the total number of self-avoiding s-t paths for the 26×26 grid graph. This is the current world record and was officially registered in the On-Line Encyclopedia of Integer Sequences [11] in November 2013. The results of big numbers are listed in Table I. The detailed techniques for solving larger problems are presented in the report by Iwashita et al. [6].

Table 3.1: The world record of the self-avoiding path enumeration problems (A007764 in [11])

n	The number of paths
1	2
2	12
3	184
4	8512
5	1262816
6	575780564
7	789360053252
8	3266598486981642
9	41044208702632496804
10	1568758030464750013214100
11	182413291514248049241470885236
12	64528039343270018963357185158482118
13	69450664761521361664274701548907358996488
14	227449714676812739631826459327989863387613323440
15	2266745568862672746374567396713098934866324885408319028
16	68745445609149931587631563132489232824587945968099457285419306
17	6344814611237963971310297540795524404944398686648069364636938785336
18	178211284084206512989384946652325257167838065704676559314524746058266922782532
19	15233449712510487999308074281031922969089994542553232945557760298667373556060592877569255844
20	3962891998230375602072995171333625021063397057394637715152371133770106823640357067044720649940398
21	3137475105013710272042053813782214513103312193698723653061359913464333739389857939655576992246021316463868
22	755970286667345339661519123315222619353103732072409481167391410479517927436312349870388833176349872711404439792
23	554354293552374700991431848906143793069037997096433132556958646484400840733488554456386924020875711242060085134829313945720
24	123717122312070647583387448626735708323730419890129435396782720808495169551159304856413945507921530371918580282125122809266003045813867 91094
25	8402974857881133471007083745436809127296054293775383549982474262393702849789812525692917857708397096012125602506027316549718402106494049 97837560424 7408
26	17369931586279272931117544042123649890037222958828814060466370372091034241327613476278921819349800610708229622314338049134829000267219311 29627708738890853908108906396

3.4 CONCLUSION

In this chapter, we presented recent research activities on BDD/ZDD-based discrete structure manipulation. Although many years have passed since BDDs/ZDDs were developed, there are still many interesting and exciting research topics related to them. Especially, the frontier-based method is important because many kinds of practical problems are efficiently solved by some variations of this algorithm.

In order to utilize such state-of-the-art techniques, our research project developed an integrated software tool set, named 'Graphillion" [5]. This tool provides an efficient means of ZDD construction by frontier-based method for a given graph problem, and also gives various ZDD operations in a simple user interface based on a graph library package written in Python. An interesting tutorial video and the open source codes are provided at the webpage of Graphillion.org. We hope that many researchers and engineers will be interested in this tool and will use it for various kinds of problems.

3.5 EXERCISES

3.1. The following graph is the same graph as Fig. 3.3 on the left but using a different variable ordering. Describe a ZDD representing the set of all s-t paths as well as Fig. 3.3 on the right.

3.2. For the graph of Fig. 3.3 on the left, draw a ZDD representing the set of all minimal cutsets between s and t. (A cutset is a set of deleting edges to make s and t disconnected.) Then, draw a BDD instead of a ZDD. Compare which is more compact and consider why.

3.3. For the graph of Fig. 3.3 on the left, draw a ZDD representing the set of all (not only minimal) cutsets between s and t. Then, draw a BDD instead of a ZDD. Compare which is more compact and consider why.

3.4. Describe all the self-avoiding paths between the opposite corners of a 2×2 grid graph.

3.5. Draw the 4th step of Fig. 3.4

3.6. For the following 2×2 grid graph, we execute the Simpath algorithm. If we use a variable ordering as $e_1, e_2, e_3, \ldots e_{12}$, how does the frontier move from s to t?

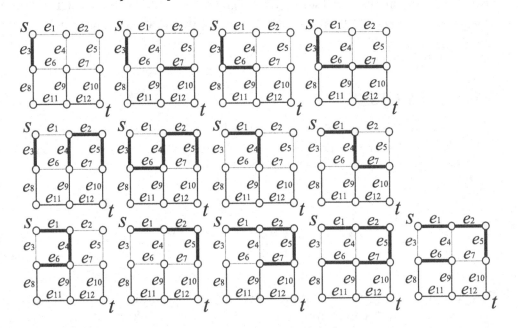

3.7. Let us consider the example shown in Fig. 3.5. After deciding the usage of edges from e_1 to e_7, we have 13 patterns as shown below. Classify the patterns into several equivalent classes based on the possible patterns of e_8 to e_{12}.

REFERENCES

[1] R. E. Bryant, "Graph-based algorithms for Boolean function manipulation," *IEEE Transactions on Computers*, Vol. C-35, No. 8, pp. 677–691, 1986. DOI: 10.1109/TC.1986.1676819. 50, 51

[2] Seiji Doi and et al, "Time with class! let's count!", 2012. YouTube video, http://www.youtube.com/watch?v=Q4gTV4r0zRs. 57

[3] Hiroshi Imai, Satoru Iwata, Kyoko Sekine, and Kensyu Yoshida, "Combinatorial and geometric approaches to counting problems on linear matroids, graphic arrangements, and partial orders," In Jin-Yi Cai and ChakKuen Wong, editors, *Computing and Combinatorics*,

Vol 1090, *Lecture Notes in Computer Science*, pp. 68–80, Springer Berlin Heidelberg, 1996. 55

[4] Takeru Inoue, Keiji Takano, Takayuki Watanabe, Jun Kawahara, Ryo Yoshinaka, Akihiro Kishimoto, Koji Tsuda, Shin-ichi Minato, and Yasuhiro Hayashi, "Distribution loss minimization with guaranteed error bound," *IEEE Transactions on Smart Grid*, Vol. 5, No. 1, pp. 102–111, 2014. DOI: 10.1109/TSG.2013.2288976. 56

[5] Takeru Inoue, et al, "Graphillion. http://graphillion.org/", 2013. 59

[6] Hiroaki Iwashita, Yoshio Nakazawa, Jun Kawahara, Takeaki Uno, and Shin-ichi Minato, "Efficient computation of the number of paths in a grid graph with minimal perfect hash functions," *Hokkaido University, Division of Computer Science, TCS Technical Reports*, TCS-TR-A-10-64, 2013. 57

[7] D. E. Knuth, *The Art of Computer Programming: Bitwise Tricks & Techniques; Binary Decision Diagrams*, Vol. 4, fascicle 1. Addison-Wesley, 2009. 50, 52, 53, 55

[8] Shin-ichi Minato, "Zero-suppressed BDDs for set manipulation in combinatorial problems," In *Proc. of 30th ACM/IEEE Design Automation Conference (DAC'93)*, pp. 272–277, 1993. DOI: 10.1145/157485.164890. 50, 51, 53

[9] S. Minato, "Recent topics on BDD/ZDD-based discrete structure manipulation," *Proceedings of the 2013 Reed-Muller Workshop* (RM-2013), May 24, 2014, Toyama, Japan, pp.1-7. 49

[10] Kyoko Sekine, Hiroshi Imai, and Seiichiro Tani, "Computing the tutte polynomial of a graph of moderate size," in John Staples, Peter Eades, Naoki Katoh, and Alistair Moffat, editors, *Algorithms and Computations*, Vol. 1004, *Lecture Notes in Computer Science*, pp. 224–233. Springer Berlin Heidelberg, 1995. DOI: 10.1007/BFb0015401. 55

[11] N. J. A. Sloane, "The on-line encyclopedia of integer sequences," http://oeis.org/. DOI: 10.1007/978-3-540-73086-6_12. 57, 58

[12] D. J. A. Welsh, Complexity: Knots, colourings and counting, *London Mathematical Society Lecture Note Series*, Vol. 186, pp. 372–390, 1993. 55

CHAPTER 4

Regular Expression Matching Using Zero-Suppressed Decision Diagrams

Shinobu Nagayama

CHAPTER SUMMARY

Regular expression matching is an operation to search for strings from an input text that match patterns described by regular expressions. Since this operation is widely used as a basic operation in many applications, such as network intrusion detection and text retrieval in databases, fast regular expression matching methods are required. Especially, in low-cost embedded systems, not only high speed but also low introduction cost, high maintainability, and small memory size are desired. This chapter focuses on fast software matching programs with small memory consumption. This chapter shows a regular expression matching method using zero-suppressed decision diagrams (ZDDs) that is intended for implementation on a software package for decision diagrams. A given regular expression is converted into a non-deterministic finite automaton (NFA), and the NFA is compactly represented using one-hot encoding and a ZDD. By realizing state transitions on the NFA as operations on the ZDD, regular expression matching is performed efficiently. Experimental results show that, by using only ZDDs, the number of nodes needed to represent NFAs can be reduced by 43.6%, on average, compared to an existing method using ordinary binary decision diagrams (BDDs) [39]. By using one-hot encoding in addition, the number of nodes can be reduced further by 6.4%, on average. This results in a reduction of memory needed for regular expression matching. In addition, the presented method using ZDDs and one-hot encoding performs regular expression matching up to about five times faster than an existing method using BDDs. In this method, such high performance can be achieved without using any hardware accelerators like FPGAs.

A preliminary version of this chapter appeared as [25].

4.1 INTRODUCTION

Regular expression matching is an operation to search for strings from an input text that match patterns described by regular expressions. For example, assume that a pattern and a text are given as follows:

- Pattern: dan (cer | ger)

- Text: hir0sieg*danger*gfa73l*dancer*go8s

Then, in this example, the pattern occurs at two positions in the text. This regular expression pattern represents a set of two string patterns: "dancer" and "danger". In this way, a regular expression can encompass tens or hundreds of string patterns [22]. Thus, regular expression matching is widely used as a basic operation in many applications. For example, it is often used when there are a number of string patterns to be searched when users want to search with ambiguous patterns that specify only some prefix characters or a suffix (e.g., "dan···" or "····.exe"). Specific examples of such applications include network intrusion detection, text retrieval in databases, and analysis of DNA and protein sequences. All examples require fast regular expression matching. Various methods for regular expression matching have been extensively studied to shorten its computation time, and a number of research results have been presented in last 50 years [2, 27].

In the 1970s and 1980s, hardware algorithms for simple character string matching [11, 23], that is a special case of regular expression matching, were presented. In addition, various hardware algorithms such as a special-purpose string matching hardware for database machines [18], content addressable memory (CAM)-based methods [43], and trie-based methods [35] have been reported. Particularly, in recent years, many hardware accelerators using FPGAs [4, 7, 12, 13, 16, 22, 26, 29, 31, 37, 38] have been developed for fast network intrusion detection systems (NIDSs). In high-end systems whose first priority is speed, such hardware accelerators are useful. However, in low-cost systems for consumers, not only speed but also *introduction cost* and *maintainability* are important, and thus, software programs are often used to satisfy these requirements. In fact, software programs are still used in modern systems [33], even though many hardware algorithms have been proposed. Thus, speeding up of software programs for regular expression matching is more useful in low-cost systems. This chapter focuses on speeding up of software programs, and shows a fast algorithm intended for software implementation without any hardware acceleration, such as an FPGA.

Software programs for regular expression matching are usually realized by deterministic or non-deterministic finite automata (DFA or NFA) [17, 22, 32, 36]. In DFAs and NFAs, regular expression matching is performed by reading a character of a given text one by one, and transferring current states to other states or the same state according to the read character. The main difference between DFAs and NFAs is the number of current states. DFAs have only one current state, while NFAs can have more than one current state. DFA-based matching methods are faster because in DFAs, only one state transition from a current state to a next state occurs for one input character, but in general, they require more memory space than NFAs. Thus, many methods [3, 9, 10, 17, 32] have been proposed to reduce the space complexity of DFAs. However, in systems that have many regular expression patterns, like NIDSs, DFA-based methods can still run out of memory due to *state-space explosion of DFAs* [39]. On the other hand, since NFAs can represent regular expression patterns *much more compactly than DFAs*, they tend to avoid state-space explosion [14], NFAs are suitable for use in such systems. However, NFA-based regular expression matching is slow. This is because in an NFA, many state transitions from a set of current states to a set of next states occur for an input character, and each state transition is processed one by one in software. Thus, to use NFAs in systems that operate at high speed, techniques to compute state transitions quickly are indispensable.

To accelerate state transitions on an NFA, hardware accelerators that can compute multiple state transitions in parallel are often used. However, an algorithmic method that achieves fast state transitions using binary decision diagrams (BDDs) has been proposed [39]. This method converts a relation of state transitions and a set of states in an NFA into logic functions by encoding states and input characters as binary numbers, and represents the logic functions using BDDs. Then, it performs state transitions on an NFA simultaneously by functional operations on those BDDs. By using functional operations on BDDs, the method achieves *two or three orders of magnitude higher performance* than a conventional NFA-based regular expression matching method. In addition, the method using BDDs requires almost the same amount of memory as the conventional method. These results are achieved due in large part to compactness of BDDs. However, BDDs do not always represent *relations* and *sets* compactly, since BDDs have been devised to represent *logic functions* compactly [6].

Thus, instead of BDDs, the method shown in this chapter uses zero-suppressed decision diagrams (ZDDs) [20], which are suitable for representations of relations and sets, to represent an NFA more compactly. It also uses one-hot encoding [30] for efficient regular expression matching. In the existing method using BDDs, since each state in an NFA is encoded as a binary number with $\lceil \log_2(n) \rceil$ bits, where n is the number of states in an NFA, a BDD is used to represent a set of states. On the other hand, in one-hot encoding, since each state is encoded as a bit, a set of states can be represented by an n-bit vector. This feature of one-hot encoding simplifies operations on ZDDs for regular expression matching. This chapter presents some operations on ZDDs for regular expression matching, and an algorithm to compute state transitions on an NFA simultaneously using those operations. Experimental results show that by using one-hot encoding and ZDDs, the presented method requires less memory and performs regular expression matching up to about five times faster than the BDD-based method [39] discussed above. This implies that the ZDD-based method is faster than conventional NFA-based methods.

This chapter is organized as follows: Section 4.2 defines regular expressions, NFAs, and ZDDs. Section 4.3 shows representations of NFA using BDDs and ZDDs, and Section 4.4 presents a regular expression matching method using ZDDs. Experimental results are shown in Section 4.5.

4.2 PRELIMINARIES

4.2.1 REGULAR EXPRESSIONS AND FINITE AUTOMATON

In this subsection, we define the regular expression, DFA, and the NFA [14].

A **regular expression** represents a finite or infinite set of strings, called a **regular set**, by a finite set of symbols. It can usually represent a regular set compactly, and thus, it is widely used to represent a large regular set, such as virus patterns for NIDSs. For example, a set of strings that begin with 1 followed by 0 or more repetitions of 01 and then followed by 0 or 1 is the regular set:

$$\{10, 11, 1010, 1011, 101010, 101011, \ldots\}$$

By using operators * and | in the regular expression, this regular set can be represented by the following regular expression:

$$1(01)^*(0|1) \,,$$

where the operator * iterates "01" 0 or more times, and the operator | chooses either 0 or 1. In the following, we use this regular expression as a running example.

A regular expression is recursively defined as follows:

Definition 2.1 *Let Σ be a finite set of characters: $\Sigma = \{a_1, a_2, \ldots, a_n\}$, R and S be regular expressions on Σ, and $L(R)$ and $L(S)$ be regular sets denoted by R and S, respectively. Then,*

1. *$\emptyset$ is a regular expression denoting the regular set $\emptyset$ (the empty set).*

2. *ϵ is a regular expression denoting the regular set $\{\epsilon\}$ (the null character).*

3. *$a_i \in \Sigma$ is a regular expression denoting the regular set $\{a_i\}$.*

4. *$R \cdot S$ is a regular expression denoting the regular set $\{rs | r \in L(R), s \in L(S)\}$. $R^1 = R$ and $R^m = R \cdot R^{m-1}$ for $m \geq 2$. The operator '$\cdot$' is usually omitted.*

5. *$R|S$ is a regular expression denoting the regular set $L(R) \cup L(S)$.*

6. *R^* is a regular expression equivalent to $\epsilon|R|R^2|\ldots$ that denotes the regular set $\{\epsilon\} \cup L(R) \cup L(R^2) \cup \ldots$, where $L(R), L(R^2), \ldots$ are regular sets denoted by $R, R^2, \ldots$, respectively.*

Regular expression matching is an operation to determine if a given string, called a text, includes a substring that matches a string in a regular set represented by a regular expression. For example, consider the text 001111 and the regular expression $1(01)^*(0|1)$. Then, only the substring 11 in the text is included in the regular set for $1(01)^*(0|1)$, and thus, a matching is achieved. In this way, the regular expression matching can be performed by checking if each substring in a text is included in a regular set.

This checking can be realized using a finite state machine (FSM) that represents the regular set, as shown in Fig. 4.1. In this case, each state in the FSM represents positions where a string

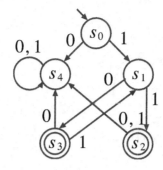

Figure 4.1: DFA for the regular expression $1(01)^*(0|1)$.

given to the FSM matches strings in the regular set from the beginning character to. The FSM reads a character of the given string one by one, and it transfers a current state, called an **active state,** to another state or the same state by the read character. When the input string matches a string in the regular set, the FSM outputs "acceptance", and a current state at that time is called an accepting state. Otherwise, it outputs "reject".

Since this FSM always has only one active state, and a next state is obtained deterministically (is uniquely specified), it is called a **deterministic finite automaton (DFA).**

Definition 2.2 *A **deterministic finite automaton (DFA)** is a 5-tuple $(Q, \Sigma, \Delta_f, q_0, A)$, where Q is a finite set of states, Σ is a finite set of characters, Δ_f is a state transition function $\Delta_f : Q \times \Sigma \to Q$, $q_0 \in Q$ is an initial state, and $A \subset Q$ is a set of accepting states. The active state is the current state $\in Q$ on the left hand side of $Q \times \Sigma \to Q$. Initially, the active state is q_0, the initial state. In a DFA, there is exactly one active state. The next state $\in Q$ on the right hand side of $Q \times \Sigma \to Q$ is determined by the active state and the present element of Σ.*

Example 2.1 *Fig. 4.1 shows a DFA for the regular expression $1(01)^*(0|1)$. In this figure, the state marked by an arrow with no state at its tail is the initial state, and the double circled states are the accepting states. The DFA transfers an active state to the next active state by traversing an arrow labeled by a character that corresponds to an input character. When a string 1011 is input one character by one character, the DFA visits $s_0, s_1, s_3, s_1,$ and s_2 in order, and outputs "acceptance" because it arrives at an accepting state.* *(End of Example)*

In a DFA, in addition to a single active state, state transitions for all characters in Σ must be defined in each state. That is, each state must have outgoing arrows for all characters. Thus, in Fig. 4.1, each state has two arrows for 0 and 1. However, it is well-known that these restrictions on a DFA can result in an excessive number of states (a condition known as **state-space explosion**). Thus, in applications that necessarily require many states, such as NIDSs, a **non-deterministic finite automaton (NFA)** is often used as an FSM. In an NFA, more than one state may be active,

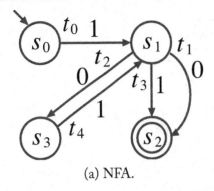

(a) NFA.

(b) State transition relation.

	x	i	y
t_0:	s_0	1	s_1
t_1:	s_1	0	s_2
t_2:	s_1	0	s_3
t_3:	s_1	1	s_2
t_4:	s_3	1	s_1

Figure 4.2: NFA for the regular expression $1(01)^*(0|1)$.

and each state does not need to have arrows for all the characters in Σ. To realize this, in an NFA, a *state transition relation* is used, while a *state transition function* is used in a DFA.

Definition 2.3 *A non-deterministic finite automaton (NFA) is a 5-tuple $(Q, \Sigma, \Delta, q_0, A)$, where Q is a finite set of states, Σ is a finite set of characters, Δ is a state transition relation $\Delta \subset Q \times \Sigma \times Q$, $q_0 \in Q$ is an initial state, and $A \subset Q$ is a set of accepting states.*

In a DFA, an active state is always transferred to another active state, since state transitions are defined as a function. On the other hand, in an NFA, an active state can transition to no state or to more than one state. If the NFA transitions to no state, the state is said to become *inactive*. If state transition relations for an input character are not defined at an active state (i.e., the state has no arrow labeled by the input character), then the state becomes inactive.

An arbitrary regular expression can be represented by an NFA as well as a DFA, and regular expression matching can be performed by state transitions on the NFA. If one of the accepting states has become active after reading all characters of an input text, then a matching is achieved (i.e., the NFA outputs "acceptance"). Otherwise, matching has failed. In many cases where matching has failed, all states in the NFA are inactive (i.e., no active state), while DFAs always have an active state. This property of NFAs could help to more quickly detect that a matching has failed.

Example 2.2 *Fig. 4.2(a) shows an NFA for the regular expression $1(01)^*(0|1)$, and Fig. 4.2(b) specifies its state transition relation $t_j = (x, i, y) \in \Delta$, where x and y denote a current state and a next state, respectively, and i denotes an input character. The NFA transfers active states by traversing arrows labeled by a character that corresponds to an input character. If an active state has no arrow labeled by the corresponding character, then the state just becomes inactive. When a string 1011 is input one character by one character, the NFA begins by transferring from the initial state s_0 to s_1, then it visits s_2 and s_3 simultaneously (note that two arrows leave state s_1, one going to s_2 and the other going to s_3). After that, s_1 and s_2 are visited in order, and a matching is achieved. Therefore, the string 1011 is accepted by the NFA.*

On the other hand, a string 1111 is not accepted because, although the second 1 brings the NFA into an accepting state s_2, the third 1 makes s_2 inactive, and thus subsequently, the NFA has no active state.   *(End of Example)*

The NFA for $1(01)^*(0|1)$ has four states. Recall that the DFA for $1(01)^*(0|1)$ has five states.

Doing Exercises 4.1 and 4.2 now would help the reader to better understand NFAs.

4.2.2   BINARY DECISION DIAGRAMS

This subsection defines decision diagrams and zero-suppressed binary decision diagrams [19, 20, 42].

Definition 2.4 *A **binary decision diagram (BDD)** [1, 6, 19] is a rooted directed acyclic graph (DAG) representing a logic function. A BDD is obtained by repeatedly applying the Shannon expansion $f = \overline{x_i} f_0 \vee x_i f_1$ to the logic function, where $f_0 = f(x_i = 0)$, and $f_1 = f(x_i = 1)$. It consists of two terminal nodes representing function values 0 and 1 respectively, and non-terminal nodes representing input variables. Each non-terminal node has two outgoing edges, 0-edge and 1-edge, that correspond to the values of the input variables. Neither terminal node has outgoing edges. When the order of the variables is the same in all paths from the root node to a terminal node, a BDD is called an ordered BDD (OBDD). In addition, it is called a reduced ordered BDD (ROBDD), when the following two reduction rules are applied:*

1. *Coalesce equivalent sub-graphs.*

2. *Delete non-terminal nodes whose two outgoing edges point to the same node v, and redirect edges, that point to the deleted node, to v.*

ROBDDs will be denoted simply as BDDs.

Definition 2.5 *A **zero-suppressed decision diagram (ZDD)** [20] is a variant of the BDD, and is obtained by applying the following two reduction rules:*

1. *Coalesce equivalent sub-graphs.*

2. *Delete non-terminal nodes whose 1-edge points to the terminal node representing 0, and redirect edges, that point to the deleted node, to the node to which the 0-edge of the deleted node has pointed.*

Example 2.3 *Figs. 4.3(a) and (b) show the BDD and the ZDD for a logic function $f_0(x_1, x_0, i, y_1, y_0)$ whose truth table is shown in Table 4.1(b). In these figures, dashed lines and solid lines denote 0-edges and 1-edges, respectively. For readability, some terminal nodes are not combined. The difference between the BDD and the ZDD comes from their reduction rules. For example, in the ZDD, when $x_1 = 0$, $x_0 = 0$, and $i = 1$, a node labeled by y_1 is deleted due to the second reduction rule. This is because the 1-edge of the deleted node has pointed to the terminal node labeled by 0. On the*

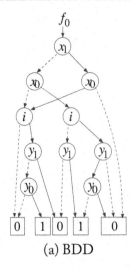
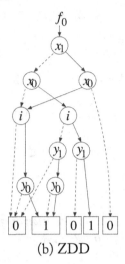

(a) BDD   (b) ZDD

Figure 4.3: BDD and ZDD for a logic function f_0.

other hand, in the BDD, the corresponding node is not deleted since the reduction rule does not specify deletion.

Note that in ZDDs, a function value cannot be obtained by only traversing a path from the root node to a terminal node, but in BDDs, it can.[1]   (End of Example)

4.3   BDDS AND ZDDS FOR NFAS

4.3.1   REPRESENTATIONS OF NFAS USING BDDS

A state transition relation of an NFA can be defined as a characteristic function f as:

$$f(x, i, y) = \begin{cases} 0 & (x, i, y) \notin \Delta \\ 1 & (x, i, y) \in \Delta \end{cases},$$

where $x \ldots$, $i \ldots$, and $y \ldots$.

This characteristic function can be converted into a logic function f_0 by encoding states and input characters in an NFA as binary numbers. Then, it can be represented using a BDD.

Example 3.1 *Table 4.1(a) shows the characteristic function f of the state transition relation in Fig. 4.2(b), and Table 4.1(b) shows its logic function f_0, where s_0, s_1, s_2, and s_3 are encoded as $(0, 0)$,*

---

[1]To find input combinations that make the function 1, in a ZDD, we have to recover the eliminated nodes. For example, in Fig 4.3(b), there is a path for $x_1 = 0, x_0 = 1, i = 1$, and $y_1 = 1$, from the root node to a constant-1 node. In this path, the node for y_0 is missing. So, we must add a node for y_0, and the edge for $y_0 = 0$. Also, there is a path for $x_1 = 0, x_0 = 0$, $i = 1$, and $y_0 = 1$, from the root node to a constant-1 node. In this path, the node for y_1 is missing. So, we must add a node for y_1, and the edge for $y_1 = 0$.

Table 4.1: Characteristic functions

(a) Characteristic function f

	s_0	0	s_0	0
	s_0	0	s_1	0
		⋮		⋮
	s_0	1	s_0	0
t_0:	s_0	1	s_1	1
	s_0	1	s_2	0
		⋮		⋮
	s_1	0	s_1	0
t_1:	s_1	0	s_2	1
t_2:	s_1	0	s_3	1
	s_1	1	s_0	0
	s_1	1	s_1	0
t_3:	s_1	1	s_2	1
	s_1	1	s_3	0
		⋮		⋮
	s_3	1	s_0	0
t_4:	s_3	1	s_1	1
	s_3	1	s_2	0
	s_3	1	s_3	0

(b) Binary encoded characteristic function f_0

	x_1	x_0	z	y_1	y_0	f_0
	0	0	0	0	0	0
	0	0	0	0	1	0
			⋮			⋮
	0	0	1	0	0	0
t_0	0	0	1	0	1	1
	0	0	1	1	0	0
			⋮			⋮
	0	1	0	0	1	0
t_1	0	1	0	1	0	1
t_2	0	1	0	1	1	1
	0	1	1	0	0	0
	0	1	1	0	1	0
t_3	0	1	1	1	0	1
	0	1	1	1	1	0
			⋮			⋮
	1	1	1	0	0	0
t_4	1	1	1	0	1	1
	1	1	1	1	0	0
	1	1	1	1	1	0

$(0, 1)$, $(1, 0)$, *and* $(1, 1)$, *respectively. Input characters can be encoded as the ASCII code for example; indeed, we do it in this project (see Table 4.5). However, in this example, they are encoded as* 0 *and* 1 *for simplicity.*

This binary encoded characteristic function can be represented using a BDD, as shown in Fig. 4.3(a). (*End of Example*)

Similarly, sets of active states (current states or next states) and accepting states in an NFA can also be represented using BDDs by defining their characteristic functions, and encoding them as binary numbers.

In this way, BDDs can represent all the information needed for state transitions on an NFA.

4.3.2 REPRESENTATIONS OF NFAS USING ZDDS

We can represent a state transition relation and sets of states using ZDDs in the same way as the previous subsection.

Example 3.2 *Fig. 4.3(b) shows the ZDD for the characteristic function defined in Table 4.1(b).* (*End of Example*)

Table 4.2: Tow binary encodings of four objects

Objects	Encodings	
	Standard binary	One-hot
O_0	00	0001
O_1	01	0010
O_2	10	0100
O_3	11	1000

In the literature [15], a method to directly represent a regular expression by a ZDD without using NFAs has been proposed for formal verification of LSI designs. As far as we know, this is the first method that has used ZDDs to represent regular expressions. Since in this method, a ZDD represents a set of strings (i.e., a regular set) defined by a given regular expression, only regular expressions with length constraints can be represented. That is, this method is intended to represent only a finite set of strings with predefined length. Thus, even the simple regular expression $1(01)^*(0|1)$ cannot be represented completely. This is because this regular expression defines an infinite set of strings with unbounded length, as shown previously. Regular expressions like this are often used to find strings with unknown length. Therefore, this representation method [15] is unsuitable for regular expression matching.

On the other hand, since in our method, ZDDs represent NFAs instead of representing regular expressions directly, our method can deal with any regular expression, and it is suitable for regular expression matching. Since ZDDs are suitable for representations of relations and sets [20], ZDDs can represent NFAs compactly. This results in efficient regular expression matching. However, the method shown in this chapter uses another way to represent states and inputs in NFAs for more efficient regular expression matching. Instead of standard binary encoding used in the existing method [39], it uses one-hot encoding to represent NFAs by ZDDs.

While the standard binary encoding encodes, for example, four objects as 00, 01, 10, and 11 using $\lceil \log_2(4) \rceil = 2$ bits, one-hot encoding encodes the four objects as 0001, 0010, 0100, and 1000 using 4 bits, as shown in Table 4.2.

Example 3.3 *By using one-hot encoding, the four states s_0, s_1, s_2, and s_3 of the NFA in Fig. 4.2(a) are represented as $(0, 0, 0, 1)$, $(0, 0, 1, 0)$, $(0, 1, 0, 0)$, and $(1, 0, 0, 0)$, respectively. By also encoding the two input characters 0 and 1 as $(0, 1)$ and $(1, 0)$, we can convert the state transition relation in Fig. 4.2(b) into the characteristic function f_1 shown in Fig. 4.4(a).*

The ZDD shown in Fig. 4.4(b) represents this characteristic function.   *(End of Example)*

Since in one-hot encoding, a binary variable corresponds to a state or an input character, it requires more binary variables than the standard binary encoding. More binary variables require more nodes, and thus, using one-hot encoding often makes the size of decision diagrams larger.

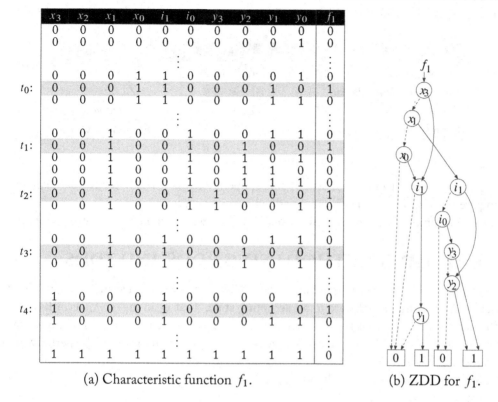

	x_3	x_2	x_1	x_0	i_1	i_0	y_3	y_2	y_1	y_0	f_1
	0	0	0	0	0	0	0	0	0	0	0
	0	0	0	0	0	0	0	0	0	1	0
						⋮					⋮
	0	0	0	1	1	0	0	0	0	1	0
t_0:	0	0	0	1	1	0	0	0	1	0	1
	0	0	0	1	1	0	0	0	1	1	0
						⋮					⋮
	0	0	1	0	0	1	0	0	1	1	0
t_1:	0	0	1	0	0	1	0	1	0	0	1
	0	0	1	0	0	1	0	1	0	1	0
	0	0	1	0	0	1	0	1	1	0	0
	0	0	1	0	0	1	0	1	1	1	0
t_2:	0	0	1	0	0	1	1	0	0	0	1
	0	0	1	0	0	1	1	0	0	1	0
						⋮					⋮
	0	0	1	0	1	0	0	0	1	1	0
t_3:	0	0	1	0	1	0	0	1	0	0	1
	0	0	1	0	1	0	0	1	0	1	0
						⋮					⋮
	1	0	0	0	1	0	0	0	0	1	0
t_4:	1	0	0	0	1	0	0	0	1	0	1
	1	0	0	0	1	0	0	0	1	1	0
						⋮					⋮
	1	1	1	1	1	1	1	1	1	1	0

(a) Characteristic function f_1.　　　　(b) ZDD for f_1.

Figure 4.4: Characteristic function by one-hot encoding and its ZDD.

However, as shown in Example 3.3, truth tables of characteristic functions obtained from state transition relations are usually sparse (i.e., densities of logic functions [24, 30] are low), and ZDDs can represent such functions compactly [20]. Unfortunately, for this small example, the ZDD of NFA with one-hot encoding is equal in size to the ZDD of NFA with the standard binary encoding. However, as shown in our experimental results, for larger functions, ZDDs of NFAs with one-hot encoding are smaller.

The reader is encouraged to do Exercises 1.5 to 1.8 here to understand better the relation between one hot encoding and ZDD oizoo.

In addition to reducing the size of ZDDs, one-hot encoding has another advantage. It results in an efficient representation of sets of states. Since in one-hot encoding, a binary variable corresponds to a state, a set of states can be represented by only a two-valued vector consisting of their binary variables, without using a decision diagram. For example, in Example 3.3, a set of current states s_2 and s_3 can be represented by a two-valued vector $(1, 1, 0, 0)$. In this way, by

Table 4.3: Comparison of representations of NFAs

	Existing method [39]	Our method
Encoding	Standard binary	One-hot
State transition relation	BDD	ZDD
Set of active states	BDD	Two-valued vector
Set of accepting states	BDD	Two-valued vector

not using any decision diagrams for sets of states, operations on a ZDD for regular expression matching are made simpler. This will be described in the next section.

Table 4.3 summarizes representations of NFAs in the existing method and our method.

4.4 MATCHING METHOD USING BDDS AND ZDDS

Recall that the process of matching is one of determining whether a substring of a given string takes the NFA from the initial state to an accepting state. That is, in an NFA, according to an input character, state transitions from current states to next states are performed. If the next states include an accepting state, then a matching is achieved. The same processes can be performed on a state transition relation Δ. The following shows the flow of regular expression matching on a state transition relation Δ. In the following, let S_a be a set of active states, and let A be a set of accepting states.

1. Let the initial state q_0 be active (i.e., the current state).

2. From the state transition relation Δ, extract elements (x, i, y), in which x is included in a set of active states S_a. That is, extract the following subset $\Delta_x \subseteq \Delta$:

$$\Delta_x = \{(x, i, y) \mid (x, i, y) \in \Delta \ and \ x \in S_a\}$$

3. From Δ_x, extract elements (x, i, y), in which i is equal to an input character c. That is, extract the following subset $\Delta_{xi} \subseteq \Delta_x$:

$$\Delta_{xi} = \{(x, i, y) \mid (x, i, y) \in \Delta_x \ and \ i = c\}$$

4. From each element (x, i, y) in Δ_{xi}, extract y, and let the set of them be a new set of active states S_a.

5. If $S_a = \emptyset$, then a match will never be found, and thus terminate the process.

6. If $S_a \cap A \neq \emptyset$, then a match has been found.

$S_a = \{s_1\}$

	x	i	y
t_0:	s_0	1	s_1
t_1:	s_1	0	s_2
t_2:	s_1	0	s_3
t_3:	s_1	1	s_2
t_4:	s_3	1	s_1

$i = 1$

	x	i	y
t_1:	s_1	0	s_2
t_2:	s_1	0	s_3
t_3:	s_1	1	s_2

(a) Step 2: Extract the current state s_1.   (b) Step 3: Narrow down by the input character 1.

	x	i	y
t_3:	s_1	1	s_2

$S_a = \{s_2\}$

(c) Step 4: Extract the next state s_2.

Figure 4.5: NFA-based regular expression matching.

7. For all input characters in a given text, 2) to 6) are iterated. If $S_a \cap A = \emptyset$ for all iterations, then a match has not been found.

For simplicity, the flow of the matching algorithm that can find only substrings beginning with the first character of a given text was shown here. But, it is easy to extend the algorithm so that it can find substrings beginning with an arbitrary character in a given text. The reader is encouraged to try this extension in Exercise 1.9.

Example 4.1 *Fig. 4.5 illustrates a part of the flow of regular expression matching on the state transition relation Δ in Fig. 4.2(b). When a set of current states is $S_a = \{s_1\}$, we extract elements from Δ as shown in Fig. 4.5(a). Then, we narrow down elements by the input character 1 as shown in Fig. 4.5(b). These operations are known as the **restriction operation** in relational algebra. In Fig. 4.5(c), the next active state is extracted from the obtained element. This operation is known as the **projection operation** in relational algebra.*   *(End of Example)*

In the next subsections, we explain matching methods according to these processes.

4.4.1   REGULAR EXPRESSION MATCHING METHOD USING BDDS [39]

Since a state transition relation and sets of states can be converted into characteristic functions, as shown in Section 4.3, the above regular expression matching on the state transition relation can be realized using *functional operations* for the characteristic functions.

Example 4.2 *The set of active states $S_a = \{s_1\}$ is represented by the characteristic function shown in Fig. 4.6(a). Note that values of the variables i, y_1, and y_0 are don't care. Thus, step 2 in the previous*

x_1	x_0	i	y_1	y_0	f_a
0	0	0	0	0	0
0	0	0	0	1	0
⋮					⋮
0	0	1	1	1	0
0	1	0	0	0	1
0	1	0	0	1	1
⋮					⋮
0	1	1	1	1	1
1	0	0	0	0	0
⋮					⋮
1	1	1	1	1	0

(a) Characteristic function f_a for $S_a = \{s_1\}$.

	x_1	x_0	i	y_1	y_0	f_2
	0	0	0	0	0	0
	0	0	0	0	1	0
	⋮					⋮
	0	1	0	0	1	0
t_1:	0	1	0	1	0	1
t_2:	0	1	0	1	1	1
	0	1	1	0	0	0
	0	1	1	0	1	0
t_3:	0	1	1	1	0	1
	0	1	1	1	1	0
	⋮					⋮
	1	1	1	1	1	0

(b) Characteristic function f_2 for Fig. 4.5(b).

Figure 4.6: Characteristic functions for $S_a = \{s_1\}$ and the table in Fig. 4.5(b).

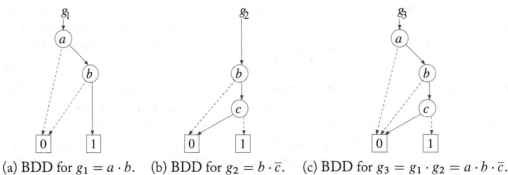

(a) BDD for $g_1 = a \cdot b$. (b) BDD for $g_2 = b \cdot \overline{c}$. (c) BDD for $g_3 = g_1 \cdot g_2 = a \cdot b \cdot \overline{c}$.

Figure 4.7: Logical AND of BDDs for g_1 and g_2.

*matching flow can be considered as the **logical AND** of the characteristic functions in Table 4.1(b) and in Fig. 4.6(a), and the resulting elements are represented by the characteristic function in Fig. 4.6(b).*
(End of Example)

In this way, we can perform regular expression matching using operations for logic functions. Such operations for logic functions can be performed efficiently on BDDs [5, 19, 39].

Example 4.3[2] *Figs. 4.7(a), (b), and (c) show BDDs for $g_1 = a \cdot b$, $g_2 = b \cdot \bar{c}$, and $g_3 = g_1 \cdot g_2$, respectively. The logical AND of g_1 and g_2 is performed by traversing their BDDs simultaneously from the root nodes. Since the BDD for g_1 has the node for a, while the BDD for g_2 has no node for a, we begin by traversing the 0-edge of the node for a. Then, we arrive at a terminal node labeled by 0 because of $g_1(a = 0) = 0$. Thus, the 0-edge of the node for a in the resulting BDD is connected to the terminal node 0. Next, we traverse the 1-edge of the node for a, and arrive at the node for b. Since the root node of the BDD for g_2 is also labeled by b, the 1-edge of the node for a in the resulting BDD is connected to a sub-BDD that represents $g_1(a = 1) \cdot g_2$. The 0-edge and the 1-edge of the node for b in the BDD for g_1 are connected to the terminal nodes 0 and 1, respectively. This means $g_1(a = 1) \cdot g_2 = b \cdot g_2$. Since $g_2 = b \cdot \bar{c}$, the sub-BDD for $g_1(a = 1) \cdot g_2$ is the BDD for g_2. Thus, the resulting BDD is obtained by connecting the 1-edge of the node for a to the BDD for g_2.*

In this way, we can perform functional operations on BDDs by traversing the BDDs in a depth first order. *(End of Example)*

The reader is encouraged to do Exercise 1.3 now to understand better the logical AND of BDDs.

By using functional operations on BDDs, we can perform regular expression matching as follows:

1. Construct three BDDs for the state transition relation Δ, a set of accepting states A, and a set of current states S_a including only the initial state q_0.

2. Compute the *logical AND* of the BDD for Δ and the BDD for S_a.

3. Let *cofactors* with respect to $i = c$ from the BDD obtained by 2).

4. Convert the BDD obtained by 3) into a *BDD that has only nodes for y*. Let the BDD be a BDD for a new set of active states S_a.

5. If the BDD for S_a represents the constant-0 function, then exit; a matching will not be achieved, and thus terminate the process.

6. Unless the *logical AND* of the BDD for S_a and the BDD for A is equal to the constant-0 function, then a matching is achieved.

7. For all input characters, 2) to 6) are iterated.

In this method, the logical ANDs of two BDDs used in 2) and 6) are realized by the *apply operation* or the *ITE operation* in a BDD package. Similarly, cofactors in 3) are obtained by the *restriction operation*, and 4) is realized by the *existential quantification* and the *variable renaming*. For more details on these operations in a BDD package, see [5, 19, 39]. Since the apply operation and the existential quantification are time-consuming operations in a BDD package, 2), 4), and 6) in this matching method account for a large part of computation time.

[2]This is not the running example.

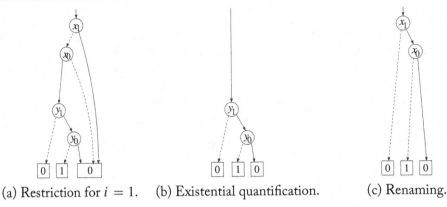

(a) Restriction for $i = 1$.   (b) Existential quantification.   (c) Renaming.

Figure 4.8: BDDs during regular expression matching.

Example 4.4 *Fig. 4.8(a) shows a BDD obtained by taking a cofactor with respect to $i = 1$ from the BDD for f_2 in Fig. 4.6(b) (i.e., by applying the restriction operation to the BDD for f_2). This operation corresponds to step 3 in the BDD-based matching method. Note that this resulting BDD has no node for i. Then, in step 4, nodes labeled by variables other than y_0 and y_1 are deleted from this BDD. The resulting BDD is shown in Fig. 4.8(b). After that, the variables y_0 and y_1 are renamed x_0 and x_1, respectively, for a new set of active states S_a. The resulting BDD is shown in Fig. 4.8(c).*
(End of Example)

In this way, by using BDDs and their operations, NFA-based regular expression matching can be performed. Computation time for conventional NFA-based matching methods depends on the number of *active states*, and thus, their time complexity is $O(|Q|)$. On the other hand, computation time for the matching method using BDDs depends on the number of *nodes*, and thus, its time complexity is $O(N_b)$, where N_b is the number of nodes. As discussed in [39], since N_b is much smaller than $|Q|$ in practical applications such as NIDSs, the method using BDDs is much faster than conventional methods.

4.4.2   REGULAR EXPRESSION MATCHING METHOD USING ZDDS

The existing method shown in the previous subsection realizes state transitions on an NFA by *functional operations* on BDDs. However, NFA-based regular expression matching are intrinsically based on *relational operations* in relational algebra, rather than functional operations, as shown in Example 4.1. Thus, using relational operations to realize the state transitions is more natural and more appropriate than using functional operations. Since the method shown in this subsection uses relational operations on ZDDs for regular expression matching, it is more suitable for regular expression matching than the existing method using BDDs. One-hot encoding allows

```
ZDDtype restriction ( ZDDtype f, vector v)
{
    IF ( f ->computed == TRUE )
        RETURN f ->result ;
    IF ( f is a terminal node )
        RETURN f ;
    IF ( f ->variable is specified by v ) {
        IF ( its value is 0 )
            f ->result = restriction ( f ->0-edge, v ) ;
        ELSE
            f ->result = union ( restriction ( f ->0-edge, v ),
                                  restriction ( f ->1-edge, v ) ) ;
    }
    ELSE
        f ->result = get_node ( f ->variable,
                                restriction ( f ->0-edge, v ),
                                restriction ( f ->1-edge, v ) ) ;
    f ->computed = TRUE ;
    RETURN f ->result ;
}
```

Figure 4.9: Pseudo-code of the restriction operation on ZDDs.

those relational operations on ZDDs. Shown below is the flow of the matching method using a ZDD for Δ and two-valued vectors for S_a, an input character c, and A.

1. Represent the initial state q_0 by a two-valued vector using one-hot encoding, and let the vector be a vector for S_a.

2. Apply the *restriction operation* to the ZDD for Δ using the vector for S_a.

3. Apply the *restriction operation* to the ZDD obtained by 2) using the vector for c.

4. Apply the *projection operation* to the ZDD obtained by 3), and let the obtained vector representing a set of next states y be a vector for new active states S_a.

5. If the vector for S_a is equal to the zero vector, then exit; a matching will never be achieved.

6. Unless the bitwise AND of the vector for S_a and the vector for A is equal to the zero vector, then a matching is achieved.

7. For all input characters, 2) to 6) are iterated.

In this method, the *restriction operation* is a relational operation in our ZDD package[3] to extract elements from Δ that include states in S_a and the input character c. Fig. 4.9 shows its

---

[3]The operation can also be implemented in a well-known open-source decision diagram package [34].

```
void projection ( ZDDtype f , a set of variables Y, vector v)
{
   IF ( f ->visited == TRUE )
      RETURN ;
   IF ( f is a terminal node )
      RETURN ;
   IF ( f ->variable is included in Y ) {
      Set the bit in v corresponding to its variable to 1 ;
   projection ( f ->0-edge, Y , v ) ;
   projection ( f ->1-edge, Y , v ) ;
   f ->visited = TRUE ;
      RETURN ;
}
```

Figure 4.10: Pseudo-code of the projection operation on ZDDs.

Table 4.4: Comparison of operations for regular expression matching

Processes	Existing method [39]	Our method
2) Specify current states	Logical AND (Apply)	*Restriction*
3) State transition	Restriction	*Restriction*
4) Specify next states	Existential quantification	*Projection*
5) Judge acceptance	Logical AND (Apply)	Bitwise AND

pseudo-code. This operation requires the root node of a ZDD and a two-valued vector for S_a or c as arguments, and returns the computed ZDD. In Fig. 4.9, *union* is an operation for union of sets, and *get_node* is an operation to make a ZDD node. Both are basic operations on ZDDs [20], and they can be performed similarly to ones on BDDs. Particularly, the union operation on ZDDs is quite similar to the logical OR on BDDs, and it also can be performed by traversing ZDDs in a depth first order, as shown in Example 4.3.

The reader is encouraged to do Exercise 1.4 here to understand better the restriction operation for ZDDs.

The *projection operation* is a relational operation in our ZDD package to extract a set of next states y from a ZDD for a state transition relation. Fig. 4.10 shows its pseudo-code. This operation requires the root node of a ZDD, a set of variables to extract, and an empty two-valued vector as arguments. It returns the two-valued vector representing a set of next states.

Example 4.5 *Fig. 4.11 shows a ZDD obtained by applying the restriction operations using vectors $(x_3, x_2, x_1, x_0) = (0, 0, 1, 0)$ for a current state $x = s_1$ and $(i_1, i_0) = (1, 0)$ for input character $i = 1$ to the ZDD in Fig. 4.4(b). These correspond to step 2 and 3 in the ZDD-based matching method.*

Figure 4.11: ZDD obtained by the restriction operations using $(x_3, x_2, x_1, x_0) = (0, 0, 1, 0)$ and $(i_1, i_0) = (1, 0)$.

Note that the resulting ZDD has no node for $x_3, x_2, x_1, x_0, i_1,$ and i_0. Then, in step 4, a vector $(y_3, y_2, y_1, y_0) = (0, 1, 0, 0)$ is extracted by applying the projection operation to this ZDD. This is because this ZDD has a non-terminal node only for y_2. The extracted vector is considered as a vector $(x_3, x_2, x_1, x_0) = (0, 1, 0, 0)$ for new active states S_a. Since the vector for the accepting state s_2 is $(0, 1, 0, 0)$, in this example, a matching is achieved. (End of Example)

In this way, the method using ZDDs of NFAs with one-hot encoding can perform regular expression matching without using the time-consuming *logical AND, existential quantification,* and *variable renaming*. Since the restriction and the projection operations in our ZDD package visit each node only once, their time complexities are $O(N_z)$, where N_z is the number of ZDD nodes.

Table 4.4 summarizes operations for regular expression matching in the existing method and our method.

4.5 EXPERIMENTAL RESULTS

To show the effectiveness of the presented matching method, we compared our matching method using ZDDs with the existing method using BDDs [39]. For our experiments, we used the following five regular expressions randomly selected from the SNORT rules [33] (the snapshot on June 20, 2013), and converted them into NFAs:

Table 4.5: Number of nodes in BDDs and ZDDs for Δ in NFAs

NFAs	Standard binary			One-hot encoding				
	BDD	ZDD	R_1 (%)	BDD	R_2 (%)	ZDD	R_3 (%)	R_4 (%)
NFA1	121	73	60.3	464	383.5	67	91.8	55.4
NFA2	133	74	55.6	515	387.2	71	95.9	53.4
NFA3	71	41	57.7	364	512.7	35	85.4	49.3
NFA4	45	24	53.3	300	666.7	20	83.3	44.4
NFA5	160	88	55.0	520	325.0	76	86.4	47.5
Average	106	60	56.4	432.6	455.0	53.8	88.6	50.0

Variable orders for decision diagrams are obtained by the sifting algorithm [28].
R_1: (ZDD with standard binary) / (BDD with standard binary) $\times 100$ (%)
R_2: (BDD with one-hot) / (BDD with standard binary) $\times 100$ (%)
R_3: (ZDD with one-hot) / (ZDD with standard binary) $\times 100$ (%)
R_4: (ZDD with one-hot) / (BDD with standard binary) $\times 100$ (%)

NFA1: /level/(0|1|2|3|4|5|6|7|8|9)$^+$/(exec|configure)

NFA2: cookies\s$^+$Monster\s$^+$server \s$^+$engine

NFA3: template\s*=\s*{$

NFA4: fn=..(/|\)

NFA5: (((\x0$b$dyndns|\x02yi) \x03org)|((\x07dynserv
         | \x04mooo)\x03com))

where \s denotes a blank character, $+$ is a regular expression operator that means $R^+ = R \cdot R^*$, and \x followed by a two-digit hexadecimal number denotes an ASCII code.

4.5.1   COMPARISON OF THE NUMBER OF NODES

Table 4.5 shows the number of nodes in four kinds of decision diagrams for Δ in NFAs: BDDs and ZDDs of NFAs with the standard binary encoding, and BDDs and ZDDs of NFAs with one-hot encoding.

From this table, we can see that by using ZDDs, the number of nodes is reduced by 43.6%, and by using one-hot encoding in addition, the number of nodes is reduced further by 6.4%, on average. The use of one-hot encoding for BDDs increases the number of nodes by 455%, on average. This is because one-hot encoding increases the number of binary variables. Thus, if we apply one-hot encoding to the existing method, we cannot improve its performance. On the other

Table 4.6: Comparison of computation times.

NFA	Computation Time (ms)		Existing / Our
	Existing	Our	
NFA1	9,890	4,160	2.4
NFA2	16,570	4,130	4.0
NFA3	11,500	3,100	3.7
NFA4	8,770	1,860	4.7
NFA5	7,040	5,060	1.4

hand, the number of nodes in ZDDs is reduced by using one-hot encoding, and thus, we can also reduce computation time for regular expression matching.

In this way, one-hot encoding not only makes operations for matching simpler, but also reduces the number of nodes in ZDDs.

4.5.2 COMPARISON OF COMPUTATION TIME

We implemented both the existing method using BDDs and our method using ZDDs on our own package, and ran them in the following computer environment: CPU: Intel Core2 Quad Q6600 2.4GHz, memory: 4GB, OS: CentOS 5.7, and C-compiler: gcc -O2 (version 4.1.2). Table 4.6 shows computation times in milliseconds of the existing method and our method needed for processing 10,000,000 input characters.

This table shows that our method is 1.4 to 4.7 times faster than the existing method for these relatively small NFAs. This is achieved due to the compactness of ZDDs and simple operations due to one-hot encoding.

4.6   CONCLUSION AND COMMENTS

This chapter presents an NFA-based regular expression matching method using ZDDs and one-hot encoding. There are two advantages of one-hot encoding: one-hot encoding makes the ZDDs smaller, and it makes operations on ZDDs simpler. That is the reason why our method is faster than the existing method using BDDs [39]. Experimental results showed that by using ZDDs, the number of nodes needed to represent NFAs is reduced by 43.6%, on average. By using one-hot encoding in addition, the number of nodes can be reduced a further by 6.4%, on average. This results in a reduction of memory needed for regular expression matching. In addition, the presented method performs regular expression matching 1.4 to 4.7 times faster than the existing method using BDDs [39]. It achieved such high performance without using any hardware accelerators.

ACKNOWLEDGMENTS

This research is partly supported by the Ministry of Education, Culture, Sports, Science, and Technology (MEXT) Grant-in-Aid for Scientific Research (C), (No. 25330071), 2013. We would like to thank Prof. Shin-ichi Minato, Mr. Ryutaro Kurai, and Mr. Takumi Makizaki for their support to this work.

4.7 EXERCISES

4.1. Show an ϵ-free NFA of the regular expression for NFA4 in Section 4.5: fn=..(/|\).

4.2. Show the state transition relation of the ϵ-free NFA obtained in the previous exercise.

4.3. Show BDDs for the characteristic functions in Fig. 4.6(a) and (b), and confirm the BDD for Fig. 4.6(b) is obtained by the logical AND of the BDD in Fig. 4.3(a) and the BDD for Fig. 4.6(a).

4.4. Show the ZDD obtained by applying the restriction operation to the ZDD in Fig. 4.4(b) using the vector for $S_a = \{s_1\}$.

4.5. The density of a logic function is given by [24]:

$$\frac{k}{2^n} \times 100(\%)$$

where k is the number of input vectors $\vec{a}$ such that $f(\vec{a}) = 1$, and n is the number of input variables. Then, compute the densities of logic functions shown in Table 4.1(b) and Fig. 4.4(a).

4.6. What does the k in the previous exercise denote as for an NFA?

4.7. Explain why one-hot encoding can reduce the number of nodes in a ZDD for a state transition relation from the viewpoint of properties of ZDDs.

4.8. Derive an upper bound on the number of non-terminal nodes in a ZDD with one-hot encoding for a state transition relation of an NFA.

4.9. Since the flows of the three matching algorithms based on NFA transitions, BDDs, and ZDDs shown in Section 4.4 can find only substrings beginning with the first character of a given text, extend the algorithms so that it can find substrings beginning with an arbitrary character in a given text.

REFERENCES

[1] S. B. Akers, "Binary decision diagrams," *IEEE Trans. Comput.*, Vol. C-27, No. 6, pp. 509–516, Jun. 1978. DOI: 10.1109/TC.1978.1675141. 69

[2] J. Aoe, *Computer Algorithms: String Pattern Matching Strategies*, IEEE Computer Society Press, 1994. 64

[3] M. Becchi and P. Crowley, "An improved algorithm to accelerate regular expression evaluation," *Proc. of the 3rd ACM/IEEE Symposium on Architecture for Networking and Communications Systems*, pp. 145–154, Dec. 2007. DOI: 10.1145/1323548.1323573. 65

[4] J. Bispo, I. Sourdis, J. M. P. Cardoso, and S. Vassiliadis, "Regular expression matching for reconfigurable packet inspection," *Proc. 2006 IEEE International Conference on Field Programmable Technology*, pp. 119–126, Dec. 2006. DOI: 10.1109/FPT.2006.270302. 64

[5] K. Brace, R. Rudell, and R. E. Bryant, "Efficient implementation of a BDD package," *Design Automation Conference*, pp. 40–45, June 1990. DOI: 10.1109/DAC.1990.114826. 76, 77

[6] R. E. Bryant, "Graph-based algorithms for boolean function manipulation," *IEEE Trans. Comput.*, Vol. C-35, No. 8, pp. 677–691, Aug. 1986. DOI: 10.1109/TC.1986.1676819. 65, 69

[7] J. Divyasree, H. Rajashekar, and K. Varghese, "Dynamically reconfigurable regular expression matching architecture," *Proc. International Conference on Application-Specific Systems, Architectures and Processors (ASAP)*, pp. 120–125, July 2008. DOI: 10.1109/ASAP.2008.4580165. 64

[8] R. Drechsler and B. Becker, *Binary Decision Diagrams: Theory and Implementation*, Kluwer Academic Publishers, 1998. DOI: 10.1007/978-1-4757-2892-7.

[9] D. Ficara, S. Giordano, G. Procissi, F. Vitucci, G. Antichi, and A. Di Pietro, "An improved DFA for fast regular expression matching," *ACM SIGCOMM Computer Communication Review*, Vol. 38, No. 5, pp. 31–40, Oct. 2008. DOI: 10.1145/1452335.1452339. 65

[10] D. Ficara, A. Di Pietro, S. Giordano, G. Procissi, F. Vitucci, and G. Antichi, "Differential encoding of DFAs for fast regular expression matching," *IEEE/ACM Transactions on Networking*, Vol. 19, No. 3, pp. 683–694, June 2011. DOI: 10.1109/TNET.2010.2089639. 65

[11] M. J. Foster and H. T. Kung, "The design of special-purpose VLSI chips," *IEEE Computer*, Vol.13, No.1, pp.26–40, 1980. DOI: 10.1109/MC.1980.1653338. 64

[12] T. Ganegedara, Y.E. Yang, and V. K. Prasanna, "Automation framework for large-scale regular expression matching on FPGA," *Proc. 2010 IEEE International Conference on Field Programmable Logic and Applications*, pp. 50–55, Sept. 2010. DOI: 10.1109/FPL.2010.21. 64

[13] J. T. L. Ho and G. G. F. Lemieux, "PERG: A scalable FPGA-based pattern-matching engine with consolidated bloomier filters," *Proc. 2008 IEEE International Conference on Field Programmable Technology*, pp.73–80, Dec. 2008. DOI: 10.1109/FPT.2008.4762368. 64

[14] J. E. Hopcroft, R. Motwani, and J. D. Ullman, *Introduction to Automata Theory, Languages, and Computation (2nd Edition)*, Addison Wesley, USA, Nov. 2000. 65, 66

[15] S. Ishihara and S. Minato, "Manipulation of regular expressions under length constraints using zero-suppressed-BDDs," *Proc. of ASP-DAC'95*, pp.391–396, 1995. DOI: 10.1145/224818.224930. 72

[16] Y. Kaneta, S. Yoshizawa, S. Minato, H. Arimura, and Y. Miyanaga, "Dynamic reconfigurable bit-parallel architecture for large-scale regular expression matching," *Proc. 2010 IEEE International Conference on Field Programmable Technology*, pp. 21–28, Dec. 2010. DOI: 10.1109/FPT.2010.5681536. 64

[17] S. Kumar, S. Dharmapurikar, F. Yu, P. Crowley, and J. Turner, "Algorithms to accelerate multiple regular expressions matching for deep packet inspection," *Proc. SIGCOMM'06*, pp. 339–350, Sept. 2006. DOI: 10.1145/1151659.1159952. 64, 65

[18] D. L. Lee and F. H. Lochovsky, "HYTREM — A hybrid text-retrieval machine for large databases," *IEEE Transactions on Computers*, Vol.39, No.1, pp.111–123, 1990. DOI: 10.1109/12.46285. 64

[19] C. Meinel and T. Theobald, *Algorithms and Data Structures in VLSI Design: OBDD – Foundations and Applications*, Springer, 1998. DOI: 10.1007/978-3-642-58940-9. 69, 76, 77

[20] S. Minato, "Zero-suppressed BDDs for set manipulation in combinatorial problems," *Proc. 30th Design Automation Conference*, pp. 272–277, 1993. DOI: 10.1145/157485.164890. 65, 69, 72, 73, 80, 98

[21] S. Minato and T. Uno, "Frequentness-transition queries for distinctive pattern mining from time-segmented databases," *Proc. 10th SIAM International Conference on Data Mining (SDM2010)*, pp. 339–349, Apr. 2010. DOI: 10.1137/1.9781611972801.30.

[22] A. Mitra, W. Najjar, and L. Bhuyan, "Compiling PCRE to FPGA for accelerating SNORT IDS," *Proc. 2007 ACM/IEEE Symposium on Architecture for Networking and Communications Systems*, pp.127–136, Dec. 2007. DOI: 10.1145/1323548.1323571. 64

[23] A. Mukhopadhyay, "Hardware algorithms for nonnumeric computation," *IEEE Transactions on Computers*, Vol.C-28, No.6, pp.384–394, 1979. DOI: 10.1109/TC.1979.1675378. 64

[24] S. Nagayama, T. Sasao, Y. Iguchi, and M. Matsuura, "Area-time complexities of multi-valued decision diagrams," *IEICE Trans. on Fundamentals*, Vol. E87-A, No. 5, pp. 1020–1028, May 2004. 73, 84

[25] S. Nagayama, "Efficient regular expression matching method using ZBDDs," *Proceedings of the 2013 Reed-Muller Workshop* (RM-2013), May 25, 2014, Toyama, Japan, pp.48-54. 63

[26] H. Nakahara, T. Sasao, and M. Matsuura, "A regular expression matching circuit based on a modular non-deterministic finite automaton with multi-character transition," *Proc. 16th Workshop on Synthesis And System Integration of Mixed Information technologies (SASIMI-2010)*, pp. 359–364, Oct. 2010. DOI: 10.1016/j.micpro.2012.05.009. 64

[27] G. Navarro and M. Raffinot, *Flexible Pattern Matching in Strings*, Cambridge University Press, 2002. 64

[28] R. Rudell, "Dynamic variable ordering for ordered binary decision diagrams," *International Conference on Computer-Aided Design (ICCAD'93)*, pp. 42–47, Nov. 1993. DOI: 10.1109/ICCAD.1993.580029. 82

[29] Y. SangKyun and L. KyuHee, "Optimization of regular expression pattern matching circuit using at-most two-hot encoding on FPGA," *Proc. 2010 IEEE International Conference on Field Programmable Logic and Applications (FPL-2010)*, pp. 40–43, Sept. 2010. DOI: 10.1109/FPL.2010.19. 64

[30] T. Sasao, *Switching Theory for Logic Synthesis*, Kluwer Academic Publishers 1999. DOI: 10.1007/978-1-4615-5139-3. 65, 73

[31] R. Sidhu and V. K. Prasanna, "Fast regular expression matching using FPGAs," *Proc. 2001 IEEE International Symposium on Field-Programmable Custom Computing Machines(FCCM-2001)*, pp. 227–238, May 2001. DOI: 10.1109/FCCM.2001.22. 64

[32] R. Smith, C. Estan, S. Jha, and S. Kong, "Deflating the big bang: fast and scalable deep packet inspection with extended finite automata," *Proc. SIGCOMM'08*, pp. 207–218, Aug. 2008. DOI: 10.1145/1402946.1402983. 64, 65

[33] Sourcefire Inc., "SNORT network intrusion detection system," http://www.snort.org/, June 20, 2013. 64, 81

[34] F. Somenzi, *CUDD Package, Release 2.5.0*, http://vlsi.colorado.edu/~fabio/CUDD/, April 2014. 79

[35] Y. Sugawara, M. Inaba, and K. Hiraki, "Over 10Gbps string matching mechanism for multi-stream packet scanning systems," *Proc. International Conference on Field Programmable Logic and Applications (FPL-2004)*, pp.484–493, Aug. 2004. DOI: 10.1007/978-3-540-30117-2_50. 64

[36] K. Thompson, "Programming technique: regular expression search algorithm," *Communications of the ACM*, Vol. 11, No. 6, pp. 419–422, June 1968. DOI: 10.1145/363347.363387. 64

[37] Y. Wakaba, M. Inagi, S. Wakabayashi, and S. Nagayama, "An efficient hardware matching engine for regular expression with nested Kleene operators," *Proc. 2011 IEEE International Conference on Field Programmable Logic and Applications (FPL-2011)*, pp. 157–161, Sept. 2011. DOI: 10.1109/FPL.2011.36. 64

[38] N. Yamagaki, R. Sidhu, and S. Kamiya, "High-speed regular expression matching engine using multi-character NFA," *Proc. 2008 IEEE International Conference on Field Programmable Logic and Applications (FPL-2008)*, pp. 131–136, Sept. 2008. DOI: 10.1109/FPL.2008.4629920. 64

[39] L. Yang, R. Karim, V. Ganapathy, and R. Smith, "Improving NFA-based signature matching using ordered binary decision diagrams," *Proc. 13th International Conference on Recent Advances in Intrusion Detection (RAID'10)*, pp. 58–78, 2010. DOI: 10.1007/978-3-642-15512-3_4. xiii, 63, 65, 72, 75, 76, 77, 78, 81, 83

[40] Y.E. Yang and V. K. Prasanna, "Space-time trade off in regular expression matching with semi-deterministic finite automata," *Proc. IEEE INFOCOM 2011*, pp. 1853–1861, April 2011. DOI: 10.1109/INFCOM.2011.5934986.

[41] Y.E. Yang and V. K. Prasanna, "Optimizing regular expression matching with SR-NFA on multi-core systems," *Proc. 2011 IEEE International Conference on Parallel Architectures and Compilation Techniques*, pp. 424–433, Oct. 2011. DOI: 10.1109/PACT.2011.73.

[42] S. N. Yanushkevich, D. M. Miller, V. P. Shmerko, and R. S. Stankovic, *Decision Diagram Techniques for Micro- and Nanoelectronic Design*, CRC Press, Taylor & Francis Group, 2006. 69

[43] S. Yusuf and W. Luk, "Bitwise optimised CAM for network intrusion detection systems," *Proc. International Conference on Field Programmable Logic and Applications (FPL-2005)*, pp.444–449, Aug. 2005. DOI: 10.1109/FPL.2005.1515762. 64

APPENDIX A

Solutions

A.1 CHAPTER 1

1.1 Both the BDD and the ZDD of a function are derived from the decision tree by applying different reduction rules. In the BDD, nodes with identical cofactors are removed. In the ZDD, nodes whose 1-cofactor points to 0-terminal are removed. In both cases, isomorphic nodes are shared. See Figure A.1.

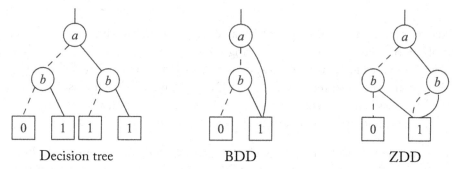

Decision tree   BDD   ZDD

Figure A.1: Problem 1.

1.2 See Figure A.2.

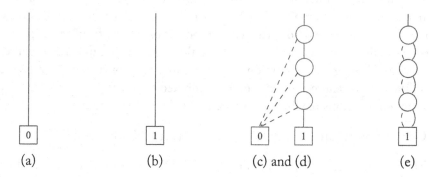

(a)   (b)   (c) and (d)   (e)

Figure A.2: Problem 2.

1.3 A ZDD representation of a function depends on as many variables as there are variables in the support of the function. A ZDD representation of the cover depends on twice as many variables, one for the positive and negative literals of each variable. See Figure A.3.

Figure A.3: Problem 3.

1.4 Each path in the BDD is disjoint with all other paths in the same BDD. That is, the product of the Boolean functions representing these two paths is the constant-0 function. This follows from the fact that there always exists a variable, which takes a different value on each pair of paths. The cover represented by a BDD is always disjoint, that is, the product of any pair of its cubes is the constant-0 function. However, in general, a cover is not disjoint. This is why we cannot represent an arbitrary cover using a standard BDD.

1.5 In unate algebra, each element is either absent or present. In binate algebra, each element is either absent or present in one of the two polarities (positive or negative). To implement operators in the binate algebra, a ZDD representation depends on twice as many variables and uses different implementation of the product and division operators. Most of the other operators, including set-union and set-difference, are the same in both cases.

1.6 Both the dot-product and the cross-product are applied to two sets of subsets. The resulting set of subsets contains all subsets derived by taking pair-wise intersections (the dot-product) or pair-wise unions (the cross-product) of subsets from the original sets. The traversal procedure performing these two operations can be the same, but the intermediate step requires choosing between union or intersection. It is convenient to implement them as two different procedures; otherwise, the computed table has to distinguish between the two operations and thus requires a third argument (in addition to the two operands).

1.7 Consider procedure $Extra\_zddLitCount()$ from Extra library [32].

1.8 The threshold operation is an important ZDD-operator that can be used, for example, in risk analysis (to remove all outcomes whose probability is less than a given threshold, assuming a uniform probability distribution) and in technology mapping (to remove non-K-feasible cuts during structural cut enumeration). Implementation is straightforward except that the threshold operation requires that we count how many times a variable appears in positive polarity on each path originating in the topmost node. If the number of appearances is more than N, the procedure returns the empty set. Otherwise, the procedure composes the result for the two cofactors.

A.2   CHAPTER 2

2.1 Note that f_2 can be represented by $f_2 = x_1 \vee x_2$.

1. The monotone increasing functions are f_2 and f_4.

2. The unate functions are f_2, f_3, and f_4.

3. The binate functions are f_1 and f_5.

2.2

$$
\begin{aligned}
f_0 &= f(x_5 = 0) = x_1 \vee x_2 \vee x_3 \vee x_4, \\
f_1 &= f(x_5 = 1) = \bar{x}_1 \vee \bar{x}_2 \vee \bar{x}_3 \vee \bar{x}_4, \\
PI(f_0) &= \{x_1, x_2, x_3, x_4\}, \\
PI(f_1) &= \{\bar{x}_1, \bar{x}_2, \bar{x}_3, \bar{x}_4\}.
\end{aligned}
$$

Also,

$$
\begin{aligned}
PI(f_0 \cdot f_1) &= PI((x_1 \vee x_2 \vee x_3 \vee x_4)(\bar{x}_1 \vee \bar{x}_2 \vee \bar{x}_3 \vee \bar{x}_4)), \\
&= \{x_1\bar{x}_2, x_1\bar{x}_3, x_1\bar{x}_4, x_2\bar{x}_3, x_2\bar{x}_4, x_2\bar{x}_1, x_3\bar{x}_4, x_3\bar{x}_1, x_3\bar{x}_2, x_4\bar{x}_1, x_4\bar{x}_2, x_4\bar{x}_3\}.
\end{aligned}
$$

Note that
$$
PI(f) \subseteq \bar{x}_5 PI(f_0) \cup x_5 PI(f_1) \cup PI(f_0 \cdot f_1).
$$

$PI(f) = \{\bar{x}_5x_1, \bar{x}_5x_2, \bar{x}_5x_3, \bar{x}_5x_4,$
$x_5x_1, x_5x_2, x_5x_3, x_5x_4,$
$x_1\bar{x}_2, x_1\bar{x}_3, x_1\bar{x}_4,$
$x_2\bar{x}_3, x_2\bar{x}_4, x_2\bar{x}_1,$
$x_3\bar{x}_4, x_3\bar{x}_1, x_3\bar{x}_2,$
$x_4\bar{x}_1, x_4\bar{x}_2, x_4\bar{x}_3\}.$

2.3 f_n has $n(n-1)$ PIs.

2.4 Five ISOPs exist for this function:

$$
\begin{aligned}
f_1 &= x_1\bar{x}_2 \vee x_2\bar{x}_3 \vee x_3\bar{x}_1, \\
f_2 &= \bar{x}_1x_2 \vee \bar{x}_2x_3 \vee \bar{x}_3x_1, \\
f_3 &= \bar{x}_1x_2 \vee \bar{x}_1x_3 \vee x_1\bar{x}_1 \vee x_1\bar{x}_2, \\
f_4 &= \bar{x}_2x_1 \vee \bar{x}_2x_3 \vee x_2\bar{x}_1 \vee x_2\bar{x}_3, \\
f_5 &= \bar{x}_3x_1 \vee \bar{x}_3x_2 \vee x_3\bar{x}_1 \vee x_3\bar{x}_2.
\end{aligned}
$$

f_1 and f_2 are MSOPs.

2.5

1. The function has 20 PIs:

$$
\begin{aligned}
&\bar{x}_1x_2,\ \bar{x}_1x_3,\ \bar{x}_1x_4,\ \bar{x}_1x_5, \\
&\bar{x}_2x_1,\ \bar{x}_2x_3,\ \bar{x}_2x_4,\ \bar{x}_2x_5, \\
&\bar{x}_3x_1,\ \bar{x}_3x_2,\ \bar{x}_3x_4,\ \bar{x}_3x_5, \\
&\bar{x}_4x_1,\ \bar{x}_4x_2,\ \bar{x}_4x_3,\ \bar{x}_4x_5, \\
&\bar{x}_5x_1,\ \bar{x}_5x_2,\ \bar{x}_5x_3,\ \bar{x}_5x_4.
\end{aligned}
$$

2. An ISOP is

$$
f = \bar{x}_1x_2 \vee \bar{x}_1x_3 \vee \bar{x}_1x_4 \vee \bar{x}_1x_5 \vee x_1\bar{x}_2 \vee x_1\bar{x}_3 \vee x_1\bar{x}_4 \vee x_1\bar{x}_5.
$$

Note that this ISOP consists of 8 PIs.

2.6 (Proof) When $x_n = 0$.
Left-hand-side of the equation: $x_1 \vee x_2 \vee \ldots \vee x_{n-1}$.
Right-hand-side of the equation: $x_1\bar{x}_2 \vee \ldots \vee x_{n-1} = x_1 \vee x_2 \vee \ldots \vee x_{n-1}$.

When $x_n = 1$.
Left-hand-side of the equation: $\bar{x}_1 \vee \bar{x}_2 \vee \ldots \vee \bar{x}_{n_1}$.
Right-hand-side of the equation: $x_1\bar{x}_2 \vee \ldots \vee x_{n-2}\bar{x}_{n-1} \vee \bar{x}_1 = \bar{x}_1 \vee \bar{x}_2 \vee \ldots \vee \bar{x}_{n-1}$.
Thus, the equality holds. □

2.7 (Proof) A PI of an n-variable function has the form

$$
x_1^{\dagger}x_2^{\dagger}\cdots x_n^{\dagger},
$$

where $x_i^{\dagger}$ denotes $\bar{x}_i$, x_i or a missing variable. Thus, at most 3^n PIs exist. □

A.3 CHAPTER 3

3.1

$\{e_1e_3e_5,\ e_1e_2,\ e_2e_3e_4,\ e_4e_5\}$

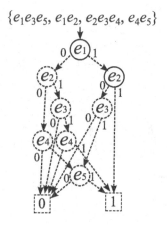

3.2

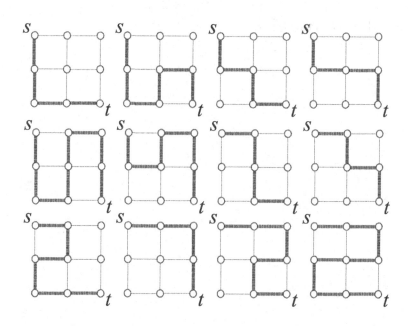

3.3

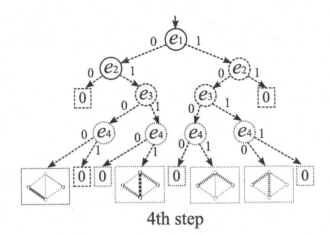

4th step

3.4 $\{\} \xrightarrow{e_1} \{s, v_2\} \xrightarrow{e_2} \{s.v_2, v_3\} \xrightarrow{e_3} \{v_2, v_3, v_4\} \xrightarrow{e_4} \{v_3, v_4, v_5\} \xrightarrow{e_5}$
$\{v_4, v_5, v_6\} \xrightarrow{e_6} \{v_4, v_5, v_6\} \xrightarrow{e_7} \{v_4, v_5, v_6\} \xrightarrow{e_8} \{v_5, v_6, v_7\} \xrightarrow{e_9}$
$\{v_6, v_7, v_8\} \xrightarrow{e_{10}} \{v_7, v_8, t\} \xrightarrow{e_{11}} \{v_8, t\} \xrightarrow{e_{12}} \{\}$

3.5

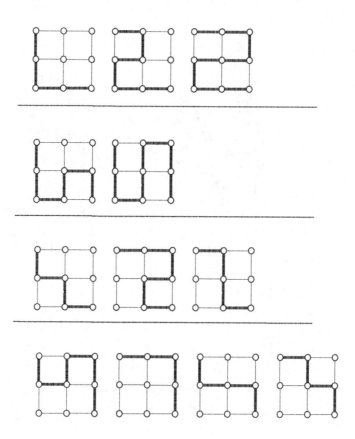

3.6

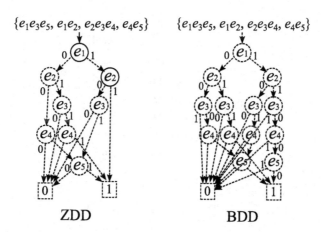

3.7 $\{e_1e_2, e_1e_2e_3, e_1e_2e_3e_4, e_1e_2e_3e_4e_5, e_1e_2e_3e_5, e_1e_2e_4, e_1e_2e_4e_5, e_1e_2e_5,$
$e_1e_3e_5, e_1e_3e_4e_5, e_2e_3e_4, e_2e_3e_4e_5, e_4e_5, e_1e_4e_5, e_2e_4e_5, e_3e_4e_5\}$

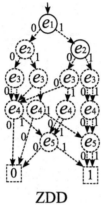 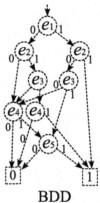

ZDD   BDD

A.4   CHAPTER 4

4.1

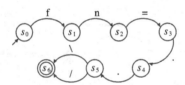

4.2

x	i	y
s_0	f	s_1
s_1	n	s_2
s_2	=	s_3
s_3	.	s_4
s_4	.	s_5
s_5	\	s_6
s_5	/	s_6

4.3

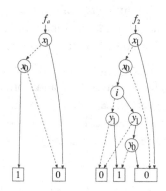

4.4

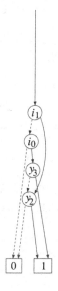

4.5 The density of the logic function f_0 is

$$\frac{5}{2^5} \times 100 = 15.625\%$$

and the density of the logic function f_1 is

$$\frac{5}{2^{10}} \times 100 = 0.4883\%$$

4.6 k denotes the number of state transitions (i.e., the number of edges) in an NFA.

4.7 The use of the one-hot encoding lowers the density significantly, as shown in Exercise 1.5, but it increases the number of input variables. However, for a state transition relation, the number of variables needed to represent each state transition (x, i, y) does not increase. Only three of the variables are needed to represent each state transition.

As shown in [20], the number of nodes in a ZDD decreases as the density decreases if the number of input variables is constant. In addition, since ZDDs automatically suppress variables for objects that never appear in any combination, the number of nodes does not increase even if such variables increase. From these two properties of ZDDs, the use of the one-hot encoding tends to reduce the number of nodes in a ZDD for a state transition relation.

4.8 From a reduction rule of ZDDs (zero-suppression), in one-hot encoding, each state transition (x, i, y) in a state transition relation is represented with three non-terminal nodes. Thus, a ZDD with the one-hot encoding for a state transition of an NFA has at most $3k$ non-terminal nodes, where k is the number of state transitions.

4.9 Regular expression matching based on NFA transitions is extended as follows. The underlined parts are extended.

1. Let the initial state q_0 be active.

2. From the state transition relation Δ, extract elements (x, i, y), in which x is included in a set of active states S_a. That is, extract the following subset $\Delta_x \subseteq \Delta$:

$$\Delta_x = \{(x, i, y) \mid (x, i, y) \in \Delta \land x \in S_a\}$$

3. From Δ_x, extract elements (x, i, y), in which i is equal to an input character c. That is, extract the following subset $\Delta_{xi} \subseteq \Delta_x$:

$$\Delta_{xi} = \{(x, i, y) \mid (x, i, y) \in \Delta_x \land i = c\}$$

4. From each element (x, i, y) in Δ_{xi}, extract y, and let the set of them be a new set of active states S_a.

5. If $S_a = \emptyset$, then $\underline{S_a = \{q_0\}}$, and skip to the step 8.

6. If $S_a \cap A \neq \emptyset$, then a match has been found.

7. $\underline{S_a = S_a \cup \{q_0\}}$.

8. For all input characters, 2) to 7) are iterated. If a matching has not been found for all iterations, then a match has failed.

Since extension of other algorithms based on BDDs and ZDDs is similar, they are omitted here.

Authors' and Editors' Biographies

ALAN MISHCHENKO

Alan Mishchenko received the M.S. degree from the Moscow Institute of Physics and Technology, Moscow, Russia, in 1993, and the Ph.D. degree from the Glushkov Institute of Cybernetics, Kiev, Ukraine, in 1997. From 1998 to 2002, he was an Intel-sponsored Visiting Scientist at Portland State University, Portland, OR. Since 2002, he has been an Associate Research Engineer in the EECS Department at University of California, Berkeley, CA. His current research interests include developing computationally efficient methods for synthesis and verification. Dr. Mishchenko was a recipient of the D.O. Pederson TCAD Best Paper Award in 2008, and the SRC Technical Excellence Award in 2011, for his work on ABC.

TSUTOMU SASAO

Tsutomu Sasao received the B.E., M.E., and Ph.D. degrees in Electronics Engineering from Osaka University, Osaka Japan, in 1972, 1974, and 1977, respectively. He has held faculty/research positions at Osaka University, Japan; IBM T. J. Watson Research Center, Yorktown Heights, NY; the Naval Postgraduate School, Monterey, CA; and Kyushu Institute of Technology, Iizuka, Japan. Now, he is a Professor at the Department of Computer Science, Meiji University, Kawasaki, Japan. His research areas include logic design and switching theory, representations of logic functions, and multiple-valued logic. He has published more than 9 books on logic design including, *Logic Synthesis and Optimization, Representation of Discrete Functions, Switching Theory for Logic Synthesis, Logic Synthesis and Verification,* and *Memory-Based Logic Synthesis,* in 1993, 1996, 1999, 2001, and 2011, respectively. He has served Program Chairman for the IEEE International Symposium on Multiple-Valued Logic (ISMVL) many times. Also, he was the Symposium Chairman of the 28th ISMVL held in Fukuoka, Japan, in 1998. He received the NIWA Memorial Award in 1979, Takeda Techno-Entrepreneurship Award in 2001, and Distinctive Contribution Awards from IEEE Computer Society MVL-TC for papers presented at ISMVLs in 1986, 1996, 2003, 2004, and 2013. He has served as an associate editor of the IEEE Transactions on Computers. He is a Fellow of the IEEE.

SHIN-ICHI MINATO

Shin-ichi Minato is a Professor at the Graduate School of Information Science and Technology, Hokkaido University. He also serves as Project Director of ERATO (Exploratory Research for Advanced Technology) MINATO Discrete Structure Manipulation System Project, executed by JST (Japan Science and Technology Agency). He received the B.E., M.E., and D.E. degrees from Kyoto University in 1988, 1990, and 1995, respectively. He had been working at NTT Laboratories from 1990 until March 2004. He was a Visiting Scholar at the Computer Science Department of Stanford University in 1997. He was a Senior Researcher of NTT Network Innovation Laboratories in 1999. He joined Hokkaido University in 2004. He started the ERATO Project in Oct. 2009. His research topics include efficient representations and manipulation algorithms for large-scale discrete structure data. He published *Binary Decision Diagrams and Applications for VLSI CAD* (Kluwer, 1995). He proposed a data structure "ZDD" (Zero-suppressed decision diagram) in 1993, which is included in the Knuth's book *The Art of Computer Programming* (Vol. 4, Fascicle 1, 2009). He is a member of IEEE, IEICE, IPSJ, and JSAI.

SHINOBU NAGAYAMA

Shinobu Nagayama received the B.S. and M.E. degrees from the Meiji University, Kanagawa, Japan, in 2000 and 2002, respectively, and the Ph.D. degree in computer science from the Kyushu Institute of Technology, Japan, in 2004. He is now an Associate Professor at Hiroshima City University, Japan. He received the Outstanding Contribution Paper Awards from the IEEE Computer Society Technical Committee on Multiple-Valued Logic (MVL-TC) in 2005 and 2013 for papers presented at the International Symposium on Multiple-Valued Logic in 2004 and 2012, respectively, the Young Author Award from the IEEE Computer Society Japan Chapter in 2009, and the Outstanding Paper Award from the Information Processing Society of Japan (IPS) in 2010 for a paper that appeared in the IPSJ Transactions on System LSI Design Methodology. His research interests include decision diagrams, regular expression matching, analysis of multi-state systems, logic design for numeric function generators, and multiple-valued logic.

JON T. BUTLER

Jon T. Butler received the B.E.E. and M.Engr. degrees from Rensselaer Polytechnic Institute, Troy, New York, in 1966 and 1967, respectively. He received the Ph.D. degree from The Ohio State University, Columbus, Ohio, in 1973. From 1987 until 2010, he was a Professor at the Naval Postgraduate School, Monterey, California. From 1974 to 1987, he was at Northwestern University, Evanston, Illinois. He is now a Distinguished Professor Emeritus. During that time he served two periods of leave at the Naval Postgraduate School, first as a National Research Council Senior Postdoctoral Associate (1980–1981) and second as the NAVALEX Chair Professor (1986–1987). He served one period of leave as a foreign visiting professor at the Kyushu Institute of Technology, Iizuka, Japan. His research interests include logic optimization, multiple-valued logic, and reconfigurable computing. He has served on the editorial boards of the IEEE Transactions on Computers, Computer, and the IEEE Computer Society Press. He has served as the editor-in-chief of Computer and the IEEE Computer Society Press. He received the Award of Excellence, the Outstanding Contributed Paper Award, and a Distinctive Contributed Paper Award for papers presented at the International Symposium on Multiple-Valued Logic. He received the Distinguished Service Award, two Meritorious Awards, and nine Certificates of Appreciation for service to the IEEE Computer Society. He is a Life Fellow of the IEEE.

Index

Printed in the United States
by Baker & Taylor Publisher Services